Mitigating Drought Impacts in Drylands

A WORLD BANK STUDY

Mitigating Drought Impacts in Drylands

Quantifying the Potential for Strengthening Crop- and Livestock-Based Livelihoods

Federica Carfagna, Raffaello Cervigni, and Pierre Fallavier, Editors

 **WORLD BANK GROUP**

Contents

Foreword

Drylands—defined here to include arid, semi-arid, and dry sub-humid zones—are at the core of Africa's development challenge. Drylands make up 43 percent of the region's land surface, account for 75 percent of the area used for agriculture, and are home to 50 percent of the population, including a disproportionate share of the poor.

Owing to complex factors, economic, social, political, and environmental vulnerability in Africa's drylands is high and rising, jeopardizing the long-term livelihood prospects of hundreds of millions of people. Climate change, which is expected to increase the frequency and severity of extreme weather events, will exacerbate this challenge. Most of the people living in drylands depend on natural resource–based livelihoods, such as herding and farming, while the ability of these activities to provide stable and adequate incomes has been eroding.

Rapid population growth has put pressure on a deteriorating resource base and created conditions under which extreme weather events, unexpected spikes in global food and fuel prices, or other exogenous shocks can easily precipitate full-blown humanitarian crises and fuel violent social conflicts. Forced to address urgent short-term needs, many households have resorted to an array of unsustainable natural resource management practices, resulting in severe land degradation, water scarcity, and biodiversity loss. African governments and the larger development community stand ready to tackle the challenges confronting dryland regions. While political will is not lacking, important questions about how the task should be addressed remain unanswered.

Do dryland environments contain sufficient resources to generate the food, employment, and income needed to support sustainable livelihoods for a fast-growing population? If not, can injections of external resources make up the deficit? Or is the carrying capacity of dryland environments so limited that outmigration should be encouraged as part of a comprehensive strategy to enhance resilience? And given the range of policy options, where should investments be focused, considering that there are many competing priorities?

To answer these questions, the World Bank teamed up with a large coalition of partners to prepare a study designed to contribute to the ongoing dialogue about measures to reduce the vulnerability and enhance the resilience of populations living in drylands. The study, based on analyses of current and projected ways to mitigate drought impacts in drylands, quantifies the potential for

strengthening crop- and livestock-based livelihoods, identifies promising interventions, quantifies their likely costs and benefits, and describes the policy
trade-offs that will need to be addressed when drylands development strategies
are devised. Sustainably developing drylands and nurturing resilience among
the people living on them will require addressing a complex web of economic,
social, political, and environmental vulnerabilities. Good adaptive responses have
the potential to generate new and better opportunities for many people, cushion
the losses for others, and smooth the transition for all. Implementation of these
responses will require effective and visionary leadership at all levels, from households to local organizations, national governments, and a coalition of development partners.

This book, one of a series of books prepared in support of the main report, is
intended to contribute to that effort.

Magda Lovei
Manager, Environment & Natural Resources Global Practice
World Bank Group

Acknowledgments

This book is one of several background works prepared for the study, "Confronting Drought in Africa's Drylands: Opportunities for Enhancing Resilience." The study, referred to here as the "Africa Drylands Study," was part of the Regional Studies Program of the World Bank Group Africa Region Vice Presidency and was conducted as a collaborative effort involving contributors from many organizations, working under the guidance of a team made up of staff from the World Bank Group and African Risk Capacity (ARC).

This series of studies aims to combine high levels of analytical rigor and policy relevance and to apply them to various topics important for the social and economic development of Sub-Saharan Africa. Quality control and oversight were provided by the Office of the Chief Economist of the Africa Region.

Raffaello Cervigni (World Bank Group) and Federica Carfagna (ARC) coordinated the overall study, working under the direction of Magda Lovei (World Bank Group).

This book, entitled *Mitigating Drought Impacts in Drylands*, was prepared by Federica Carfagna and Raffaello Cervigni based on input received from contributors, including Jawoo Koo, Frank Place, and Hua Xie (International Food Policy Research Institute, IFPRI), and Cornelis de Haan and Pierre Fallavier (World Bank consultants).

Funding for the report was provided by the Nordic Development Fund and the World Bank Group Africa Regional Studies Program.

About the Editors and Authors

About the Editors

Federica Carfagna is a statistician and vulnerability analyst for African Risk Capacity (ARC). She has been with ARC since its inception in 2009 and is one of the main authors of the methodology underlying the Africa RiskView software, the technical engine of ARC, to model the impact of drought on vulnerable populations and create the country-specific risk profiles as a basis for ARC insurance. She holds a master's degree in statistics from the University of Rome "La Sapienza" and spent one year in an exchange program at the Cass Business School in London. Before joining ARC, she worked as a statistician for the World Food Programme (WFP), the United Nations Department of Economic and Social Affairs in New York, the International Fund for Agricultural Development, and Rome City Hall. She has also worked as a data analyst for WFP's World Hunger Series and for many school-feeding publications.

Raffaello Cervigni is a lead environmental economist with the Africa Region of the World Bank. He holds master's and PhD degrees in economics from Oxford University and University College, London, and has 19 years of professional experience in programs, projects, and research financed by the World Bank, the Global Environment Facility, the European Union, and the government of Italy in a variety of sectors. He is the World Bank's regional coordinator for climate change in the Africa Region, after serving for about three years in a similar role for the Middle East and North Africa Region. He is the author or coauthor of more than 40 technical papers and publications, including books, book chapters, and articles in academic journals.

Pierre Fallavier is a social scientist and planner with 20 years of experience in social development and humanitarian programs and policies in Asia and Africa, working with aid agencies, local governments, civil society, and academia. He designed, led, and evaluated development projects and responses to crises in post-conflict and fragile areas; formulated and assessed social protection, poverty reduction, and disaster-risk management initiatives; and led policy research. He holds a master's degree in community planning from the University of British Columbia and a PhD in urban studies from the Massachusetts Institute of Technology. He currently works as chief of Social Policy, Planning, Monitoring and Evaluation for UNICEF South Sudan.

About the Authors

Cornelis (Cees) de Haan is a retired World Bank senior adviser. He has a post-graduate degree in animal production from the University of Wageningen, the Netherlands. He worked for 10 years in Dutch Technical Assistance Programs in rural development projects in South America, followed by seven years in live-stock research at the International Livestock Center for Africa (now ILRI) in Addis Ababa, Ethiopia, ultimately as deputy director general. He joined the World Bank in 1983, working for 10 years on livestock development in the African and East European regions. For the same period, he worked as adviser (later senior adviser) responsible for the policy development and quality enhancement of the World Bank's animal resource development activities in the Rural Development Department, contributing to World Bank policy and invest-ments in livestock-related environmental, health, and social issues. Since his retirement in 2001, he has remained active as a consultant in animal production and health for the World Bank and other international organizations. Within the World Bank, he contributed to several rural strategy and sector papers. He has published extensively through both World Bank publications and the scientific literature on private and public sector roles in the delivery of animal health and production services (including One Health), livestock and environment, pasto-ralism, food safety, and livestock value chains.

Jawoo Koo is a research fellow at the Environment and Production Technology Division of the International Food Policy Research Institute (IFPRI). He holds master's and PhD degrees in agricultural and biological engineering from the University of Florida. He has more than 10 years' of experience in the develop-ment of a large-scale, spatially explicit crop system modeling framework and its application in Sub-Saharan Africa. He serves as the leader of IFPRI's Spatial Data and Analytics Theme. He is the author of more than 20 technical papers and publications, including books, book chapters, and articles in academic journals.

Michael Morris is a lead agricultural economist with the Agriculture Global Practice of the World Bank. He holds master's and PhD degrees in agricultural economics from Michigan State University. He coauthored World Bank flagship publications on fertilizer policy and agricultural commercialization, and he con-tributed to *World Development Report 2008: Agriculture for Development*. His areas of expertise include agricultural policy, farm-level productivity enhancement, marketing systems and value chain development, agricultural research and tech-nology transfer, innovation systems support, institutional strengthening, and capacity building. Before joining the World Bank, he spent 16 years in Mexico, Thailand, and Washington, DC, with the International Maize and Wheat Improvement Center (CIMMYT) and IFPRI.

Frank Place is a senior research fellow with the Policies, Institutions, and Markets Program (PIM) hosted by IFPRI, where he leads research on technology adoption and impact assessment. He holds master's and PhD degrees in economics from the University of Wisconsin-Madison. Before joining PIM, he worked for more than 15 years for the World Agroforestry Centre in Nairobi. He conducted many studies related to policy constraints to and impacts of agroforestry practices. Previously, he worked for the Land Tenure Center and the World Bank, conducting studies of indigenous tenure systems in Africa.

Joanna Syroka is a cofounder and the director of research and development for African Risk Capacity (ARC). In these roles, she oversaw the ARC design phase work program and now leads the agency's technical and research work. Before joining ARC, she worked with the World Bank and the World Food Programme to develop tailored weather and commodity risk management products for agricultural and humanitarian applications in Africa, Asia, and Central and South America. Her work led to the first sovereign-level weather derivative products in Africa and the early farmer weather insurance transactions in India. Previously, she worked as a commodity derivatives analyst for one of the United Kingdom's largest utility companies. She holds a PhD in atmospheric physics from Imperial College, London.

Hua Xie is a research fellow at IFPRI. He holds a PhD in environmental engineering from the University of Illinois at Urbana-Champaign. His area of expertise is water resources and environmental system analysis and modeling. At IFPRI, his research focuses on developing quantitative analytical and modeling tools to inform policy making for sustainable management of water and other natural resources key to agricultural development. Research topics of interest include climate change impact on agricultural water resources, long-term projection of agricultural nutrient pollution, and evaluation of water and land management technologies. He has been involved in a series of studies on irrigation investment potential in Sub-Saharan African countries at both the regional and national levels.

Abbreviations

ARC	African Risk Capacity
ARV	Africa RiskView
BAU	business as usual
B/C	benefit/cost assessment
CIRAD	Agricultural Research for Development
DSSAT	Decision Support System for Agrotechnology Transfer
FAO	Food and Agriculture Organization of the United Nations
FMNR	farmer-managed natural regeneration
GDP	gross domestic product
GEPR	growth elasticity of poverty reduction
GIS	geographic information system
GLEAM	Global Livestock Environmental Assessment Model
ha	hectare
IFPRI	International Food Policy Research Institute
IIASA	International Institute for Applied Systems Analysis
IMPACT	International Model for Policy Analysis for Agricultural Commodities and Trade
IRR	internal rate of return
LSI	large-scale irrigation
SHIP	Survey-based Harmonized Indicators Program
SSI	small-scale irrigation
TLU	tropical livestock unit
UN	United Nations
WFP	World Food Programme of the United Nations
WRSI	Water Requirement Satisfaction Index

Overview

Background

"Drylands" are defined as regions having an Aridity Index of 0.65 or less. They account for three-quarters of Sub-Saharan Africa's cropland, two-thirds of its cereal production, and four-fifths of its livestock holdings.

Today, frequent and severe shocks, especially droughts, limit the livelihood opportunities available to millions of households and undermine efforts to eradicate poverty in the drylands. These shocks regularly cause large drains on government budgets and consume a significant portion of the region's international development assistance, especially in the absence of robust social protection systems and rapidly scalable safety nets. As a result, scarce resources are diverted from pursuing longer-term development goals and redirected to mobilizing costly, short-term responses to humanitarian crises.

If the current situation is precarious, the future promises to be even more challenging. By 2030, the number of people living in the drylands of East and West Africa is expected to increase by 65–80 percent (depending on the fertility scenario). Over the same period, climate change could expand the area classified as drylands by as much as 20 percent under some scenarios, for the region as a whole, with much larger increases in some countries.

Prospects for sustainable development of drylands are assessed in this book through the lens of *resilience*, understood here to mean the ability of people to withstand and respond to droughts and other shocks. Resilience is affected by three types of factors: exposure, sensitivity, and coping capacity. Other conditions being constant, a household's resilience in the face of droughts and other shocks increases the lower its exposure, the lower its sensitivity, and the greater its coping capacity.

The Model

An original modeling framework developed expressly for this analysis provides a common analytical framework for integrating findings emerging from background analyses carried out in different sectors. In four main steps, the model did the following:

- Estimated a 2010 baseline of the number of vulnerable people living in drylands
- Estimated how the number of vulnerable people living in drylands might evolve by 2030 under a range of assumptions concerning the main demographic scenarios (low-, medium-, and high-fertility) and economic drivers of change (for example, projected gross domestic product [GDP] growth)
- Calculated the number of people affected, starting from the vulnerability of households in the different administrative units of each country
- Evaluated different types of interventions in agriculture and livestock for mitigating drought impact by (a) calculating the potential for reducing the number of people affected under each scenario and (b) conducting a simplified benefit/cost (B/C) analysis for each type of intervention.

The model thus provided a coherent albeit simplified analytical framework to anticipate the scale of the challenges likely to arise in drylands, as well as to generate insights into opportunities for addressing those challenges.

Data availability defined the coverage of the overall resilience model. Coverage was limited to the 10 focus countries in the livestock models for which all the necessary data were available: Burkina Faso, Chad, Ethiopia, Kenya, Mali, Mauritania, Niger, Nigeria (the northern part), Senegal, and Uganda. The countries included in the overall resilience analysis account for 85 percent of the projected 2030 population in West Africa and nearly 70 percent of that year's population in East Africa.

The Interventions

Vulnerability profiles define the percentages of the population living in each area that are likely to be affected by mild, medium, and severe drought when they occur and the percentage of the population not at risk of drought at all.

For livestock, a series of simulation models were used to estimate the likely impacts of resilience-enhancing interventions on feed balance, livestock production, and household income resilience under different climate scenarios (baseline, mild drought, and severe drought). For agriculture, the DSSAT (Decision Support System for Agrotechnology Transfer) framework was used to assess the potential impact on yields likely to result from adoption of four best-bet crop-farming technologies (drought-tolerant varieties, heat-tolerant varieties, additional fertilizer, water harvesting techniques), agroforestry practices, irrigation, and selected combinations thereof.

The Results

Across the 10 dryland countries for which sufficient data were available, in 2010 an estimated 30 percent of the population living in dryland zones was vulnerable to droughts and other shocks. But among the people who are exposed, sensitive, and unable to cope (hence vulnerable) in any given year, only some will actually experience a drought. The frequency, geographical scale, and severity of shocks are stochastic and will vary considerably from year to year.

Assuming historical climate patterns, the modeling simulations show that in any given year the average share of people living in drylands expected to be affected by drought, depending on the country, ranges from 7 to 20 percent, with an overall average of 14 percent.

With the exception of Burkina Faso, by 2030 all countries in the sample are projected to experience increases in the number of vulnerable and drought-affected people. The projected increases reflect the combined effects of several key drivers, including rapid population growth, relatively slow and inequitable economic growth, and binding bioclimatic and social constraints that limit the natural resource base's ability to support greater numbers of animals. To put the magnitude of the resulting challenge in perspective, the annual cost of bringing the income of all drought-affected people above the poverty line by providing support through social safety nets would range from 0.3 percent to almost 5 percent of GDP.

Policy Implications

It will be prohibitively expensive for governments in dryland countries to rely on social safety nets to protect vulnerable households from the adverse effects of droughts and other shocks. Given this, policy makers will want to know the extent to which the coming challenge can be mitigated by making current liveli-hood strategies more resilient. The model results suggest that by improving the productivity of livestock and crop-farming systems in the drylands and enhancing resource access (to grazing water for livestock, for example), the interventions could considerably slow the projected increase in the number of drought-affected people.

In particular, the model shows that without the interventions, by 2030 the number of drought-affected people is projected to increase by 60 percent compared to 2010. With the interventions, the projection is an increase of only 27 percent (an improvement of 43 percentage points). In some countries, notably Ethiopia and to a lesser extent Kenya and Nigeria, by 2030 the adoption of improved management of livestock and crop-farming systems could reduce the absolute number of drought-affected people relative to the 2010 baseline. In other countries, particularly Niger but also Senegal and Mauritania, the best-bet interventions would have a more modest impact, and the number of drought-affected people in 2030 would still be considerably larger than in 2010.

Cost Implications of Interventions

How cost-effective the interventions are when compared with alternative strategies for reducing vulnerability and increasing resilience in the drylands is answered by a simple B/C assessment. The results suggest that the benefits—expressed in terms of reduced cash transfers needed to support drought-affected people—far exceed the costs of implementing the best-bet interventions. In most countries (except Mauritania and Niger), the results are robust under a wide range of cost assumptions: even if costs increase fourfold, the B/C ratio remains well above 1.

CHAPTER 1

Introduction

Defining "Drylands"

What exactly are "drylands"? While the term is commonly used, it has different interpretations. For reasons of simplicity, and consistent with widespread practice, in this book *drylands* are defined on the basis of the Aridity Index. Under this approach, which was endorsed by the 195 parties to the United Nations Convention to Combat Desertification and is also used by the United Nations Food and Agriculture Organization, drylands are defined as regions having an Aridity Index of 0.65 or less (UNEP 1997).

Aridity is usually expressed as a generalized function of precipitation, temperature, and/or potential evapotranspiration (PET). The Aridity Index can be used to quantify precipitation availability over atmospheric water demand and is calculated[1] as follows:

$$\text{Aridity Index} = MAP / MAE$$

where MAP = mean annual precipitation and MAE = mean annual PET.

Drylands can be subdivided into four aridity zones as shown in table 1.1. Because the hyper-arid zone is incapable of supporting crop and livestock production activities, it is very sparsely populated, making it of little interest to policy makers. For purposes of this book, therefore, drylands are defined as areas characterized by an Aridity Index between 0.05 and 0.65, encompassing the arid zone, the semi-arid (both dry and wet) zones, and the dry sub-humid zone.

Drylands account for three-quarters of Sub-Saharan Africa's cropland, two-thirds of cereal production, and four-fifths of livestock holdings (map 1.1). In East and West Africa—the focus of this analysis—drylands are home to almost 248 million people, and they account for a large share of the poor, including many of those lacking access to basic services such as health care and education.

Today, frequent and severe shocks, especially droughts, limit the livelihood opportunities available to millions of households and undermine efforts to eradicate poverty in the drylands. These shocks regularly cause large drains on government budgets and consume a significant portion of the region's international

Table 1.1 Aridity Zones Defined

Aridity class	Definition	Aridity Index range
1–2	Hyper-arid	0.00–0.05
3	Arid	0.05–0.20
4	Dry semi-arid	0.20–0.35
5	Wet semi-arid	0.35–0.50
6	Dry sub-humid	0.50–0.65

Source: UNEP 1997.

Map 1.1 Dryland Regions of West and East Africa

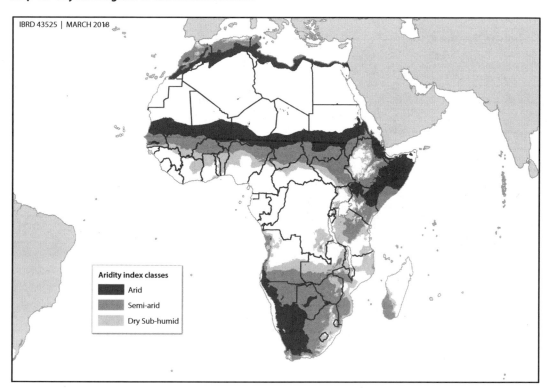

Source: ©Harvest Choice 2015. Reproduced, with permission from Zhe Guo; further permission required for reuse.

development assistance, especially in the absence of robust social protection systems and rapidly scalable safety nets. As a result, scarce resources are diverted from pursuing longer-term development goals and redirected to mobilizing costly, short-term responses to humanitarian crises. In 2011 around US$4 billion was spent on humanitarian assistance to the Sahel and the Horn of Africa, equivalent to over 10 percent of total Official Development Assistance to all of Sub-Saharan Africa (OECD 2015). The challenges threatening the livelihoods of

Map 1.2 Shift and Expansion by 2050 of Dryland Areas Caused by Climate Change

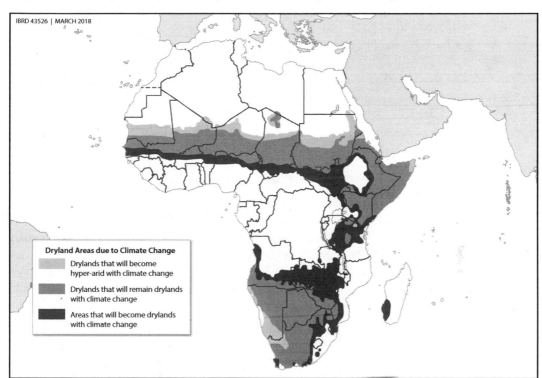

Source: Estimates based on Intergovernmental Panel on Climate Change (IPCC) data.

many of the groups that live in drylands are compounded by those groups' social and political marginalization, which muffles their voices and limits their ability to influence political processes that affect their well-being.

If the current situation is precarious, the future promises to be even more challenging. By 2030, the number of people living in the drylands of East and West Africa is expected to increase by 65–80 percent, depending on the fertility scenario, as shown in map 1.2. Over the same period, climate change could expand the area classified as drylands by as much as 20 percent under some scenarios for the region as a whole, with much larger increases in some countries. This expansion would bring more people into an ever more challenging environment.

Definition of "Resilience"

Prospects for sustainable development of drylands are assessed in this book through the lens of resilience. But what exactly is meant by *resilience?* Most definitions of resilience relate to the ability of people or ecosystems, or both, to withstand and recover from shocks. In the context of drylands, the most important of these are meteorological shocks, especially droughts, the main focus of the discussion that follows. Other shocks that are considered but not analyzed in detail include health shocks, price shocks, and conflict-related shocks. In the absence of a single, widely accepted definition of resilience, this book uses a dimension-based approach. Resilience—understood here to mean the ability of people to withstand and respond to droughts and other shocks—is affected by three types of factors: *exposure, sensitivity,* and *coping capacity.*

Other conditions being constant, a household's resilience in the face of droughts and other shocks increases the lower its exposure, the lower its sensitivity, and the greater its coping capacity. Resilience is determined by the interplay of all three dimensions, so attempts to understand resilience in terms of just one or two dimensions can produce a misleading picture. For example, when relatively few people are living below the poverty line, it would be easy to conclude that the coping capacity of the population is relatively high, since most households have enough assets to recover from a drought, should one occur. On the basis of such reasoning, policy makers might use the poverty head count as an indicator of vulnerability. But focusing on this single dimension of resilience could lead policy makers to overlook the fact that even though most households have enough assets to recover from a drought, the livelihood strategy that allowed them to accumulate those assets may be extremely sensitive to droughts. If this is the case, recurrent droughts could cause households to move in and out of poverty over time. In such a scenario, the population at risk should be understood to include not only the people who are poor today, but also the people who risk becoming poor tomorrow because their income is sensitive to droughts.

The importance of using a multidimensional approach to understand resilience is illustrated by the experience of several thousand Ethiopian households that participated in a series of surveys carried out during the period 1994–2009. Many of these households transitioned in and out of poverty, so even during a period when Ethiopia's overall poverty headcount was gradually coming down, the fortunes of individual households were much more variable. As table 1.2 shows, on average in any given year, 16–17 percent of households started out poor and stayed poor, 18–19 percent of households started out non-poor and fell into poverty, 16–20 percent of households started out poor and climbed out of poverty, and 45–48 percent of households started out non-poor and remained non-poor (Scandizzo et al. 2014).

Table 1.2 Shares of Households in Transition across Poverty Status, Ethiopia, 1994–2009

Year	Moved into poverty (%)	Stayed poor (%)	Stayed non-poor (%)	Moved out of poverty (%)
1999	18	17	45	20
2004	19	16	48	16
2009	18	17	46	19

Source: Scandizzo et al. 2014.

Figure 1.1 Poverty Head Count by Aridity Zone, Selected East and West African Countries, 2010

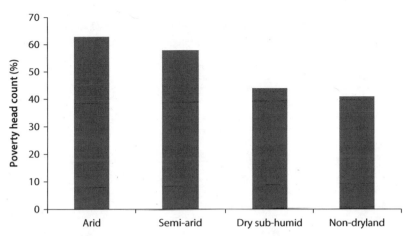

Source: D'Errico and Zezza 2015.
Note: Based on data collected in selected countries with significant drylands: Ethiopia, Malawi, Niger, Nigeria, Tanzania, and Uganda.

Ethiopia's household-level evidence generates two important insights. First, policies that succeed in bringing some people out of poverty at a particular point in time do not necessarily guarantee that many of those people will not fall back into poverty as a result of subsequent shocks. Second, enhanced resilience is a precondition for sustained reduction and eventually eradication of poverty. As a result, it makes sense to explore policies and interventions that can increase resilience (as these will lay the foundation for poverty reduction); these policies and interventions should holistically address all three dimensions of resilience.

As seen in figure 1.1, in selected countries with significant drylands the largest share of poor households (over 60 percent) live in arid lands, followed by semi-arid lands (about 58 percent), and then by dry sub-humid and non-dryland areas, where more than half of the population has the means to withstand a shock.

Note

1. Global mapping of mean Aridity Index from 1950–2000 at 30 arc-second spatial resolution.

References

D'Errico, M., and A. Zezza. 2015. *Livelihoods, Vulnerability, and Resilience in Africa's Drylands: A Profile Based on the Living Standards Measurement Study-Integrated Surveys on Agriculture*. Unpublished report. Washington, DC: World Bank.

HarvestChoice. 2015. "Dryland Regions of West and East Africa." International Food Policy Research Institute, Washington, DC., and University of Minnesota, St. Paul.

OECD (Organisation for Economic Co-operation and Development). 2015. "International Development Statistics (IDS) Online Databases." OECD, Geneva. http://www.oecd .org/dac/stats/idsonline.htm.

Scandizzo, P. L., S. Savastano, F. Alfani, and A. Paolantonio. 2014. *Household Resilience and Participation in Markets: Evidence from Ethiopia Panel Data*. World Bank Studies. Washington, DC: World Bank.

UNEP (United Nations Environment Programme). 1997. "World Atlas of Desertification 2ED." UNEP, London.

CHAPTER 2

Methodology

An original modeling framework developed for this specific analysis provides a common analytical framework for integrating findings emerging from different sectors. The model included four main steps.

First, the model estimated a 2010 baseline of the number of vulnerable people living in drylands. Vulnerable people, as previously defined, are those who are

- Exposed to drought
- Sensitive to drought, and
- Unable to cope with the impacts of drought.

Second, the model estimated how the number of vulnerable people living in drylands might evolve by 2030, under a range of assumptions concerning the main demographic and economic drivers of change. Various scenarios were developed based on three sets of population projections developed by the United Nations under low-, medium-, and high-fertility scenarios, disaggregated by country and aridity zone. These were combined with different scenarios of projected gross domestic product (GDP) growth and their resulting effects on agricultural employment patterns and poverty rates.

Third, the model calculated the number of people affected, starting from the vulnerability of households in the different administrative units and aridity zones of each country.

Fourth, the model evaluated different types of interventions in agriculture and livestock for mitigating drought impact by calculating the potential for reducing the number of people affected for each scenario.

For every type of intervention, a simplified benefit/cost (B/C) analysis was conducted.

The model thus provides a coherent, albeit simplified, analytical framework that can be used to anticipate the scale of the challenges likely to arise in drylands, as well as to generate insights into the opportunities for addressing those challenges.

Country Coverage

Because the various analyses required different types of information, coverage varied depending on data availability. The data required for the overall population projections were available for all or almost all countries in Sub-Saharan Africa, but those for the vulnerability analysis were not available for all countries. For East and West Africa, the two subregions on which the analysis concentrated, coverage was quite limited for East Africa but much more complete for West Africa. The data required for the resilience analysis similarly were not available for all countries, although here the extent of coverage varied depending on the intervention, as follows (map 2.1):

- Irrigation development: Data were available for all countries.
- Rainfed cropping systems: Data were available for most of the countries classified as dryland countries.
- Livestock systems: Data were available for a subset of dryland countries.

The coverage of the overall resilience modeling analysis was defined by the livestock systems model, which had the narrowest coverage. While data to calculate vulnerability profiles for 2010 were available for all countries in West Africa and for 61 percent of East African countries, countries included in the overall resilience analysis accounted for 85 percent of the projected 2030 population in West Africa and nearly 70 percent of the projected population in East Africa (figure 2.1).

Map 2.1 Data Availability by Type of Intervention

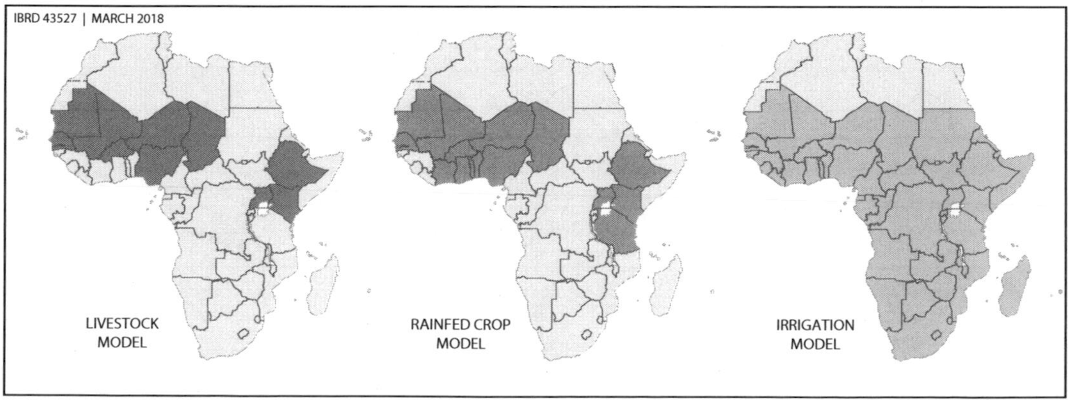

Source: African Risk Capacity 2015.

Figure 2.1 Model Coverage: Drylands Population Equivalents for Countries Included in the Analysis

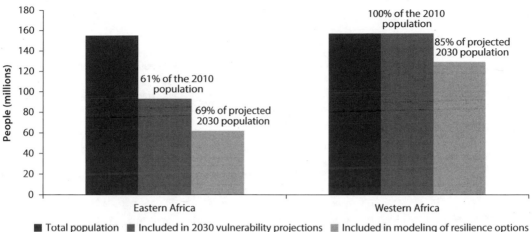

Source: World Bank calculations based on population data from United Nations Population Fund.

Main Assumptions and Sources of Data

The numbers of vulnerable and drought-affected people were estimated as follows:

$E = \alpha P$ Exposed people (E) are people living in drylands, a fraction of the country population (P).

$S = \beta E$ Sensitive people (S) are a fraction of exposed people (the agriculture-dependent).

$V = \gamma S$ Vulnerable people (V) are a fraction of sensitive people (people below given poverty lines).

$\gamma = f(U)$ The fraction (γ) is a function of people who are unable to cope with shocks (U) since they are below the poverty line.

$D = \delta V$ Drought-affected people (D) are those who are vulnerable *and* actually experience a drought in any given year.

Relationship between Resilience and Poverty

What is the relationship between resilience and poverty? Poverty reduction remains a high-order objective of development policy; building resilience to shocks is not necessarily a goal in itself, but it is an essential precondition for achieving poverty reduction. The reason is that when households and communities are repeatedly hit by shocks and lack the means to respond, they have difficulty accumulating the human, physical, and natural capital needed to lift themselves out of poverty. Increasing resilience will not automatically lead to poverty reduction; for poverty to be reduced, a number of additional actions must be taken; for example, improving health services, strengthening educational systems, and improving access to markets for inputs and outputs.

Mitigating Drought Impacts in Drylands • http://dx.doi.org/10.1596/978-1-4648-1226-2

Table 2.1 Approach and Assumptions Used in Estimating Vulnerability in the Model

Dimension of vulnerability	Proxy metric	Methodology of estimation	Assumptions
Exposure (E)	Number of people living in drylands	Spatialized United Nations median population projections.	For each country, drylands population growth rates are assumed to be equal to the national average.
Sensitivity (S)	Number of people in drylands dependent on agriculture	For 2010, country estimates are from Fox et al. (2013); for 2030, country estimates are obtained from best fit of inverse relationship between per capita income and share of employment in agriculture; for each country, employment in agriculture by 2030 is estimated by applying the regression coefficient to the income per capita projected by 2030.	The cross-country regression coefficient explaining share of employment as a function of GDP is assumed to remain valid for the next 20 years; the share of the working-age population in the total population is assumed constant and so is the share of employment in the total working-age population.
Vulnerability to mild, medium, and severe drought (V)	Number of sensitive people that are unable to cope with drought shocks (U)	For a regular year, people exposed who are also sensitive and below the poverty line are considered vulnerable. In a year of mild drought, all those with an income no more than 15% over the poverty line (US$1.44 per day) In a year of medium drought, all those with an income no more than 30% over the poverty line (US$1.63 per day) are considered vulnerable. In a year of severe drought, all those with an income no more than 45% above the poverty line (US$1.81 per day). are considered vulnerable. For livestock, where interannual and carryover effects are very important, the resilience level was set as the average of US$1.25/day/capita over the period 2012–2030.	Mild drought decreases households' income by 15%, medium drought by 30%, and severe drought by 45%. In livestock model, feed availability from BIOGENERATOR model at 60% of baseline in mild drought and 50% of baseline in severe drought scenario.
Drought affected (D)	Number of vulnerable people living in drought-affected areas	The seasonal drought index is compared to its benchmark; depending on the deviation from normality, a certain level of drought severity is triggered. The vulnerability profile of the country is used to determine how many people are affected by the drought detected.	Droughts occurring in an administrative unit homogenously affect all people living in that area.

Source: African Risk Capacity Agency 2015.

But even if increasing resilience is not a sufficient condition for poverty reduction, it is a necessary one, because households that are unable to cope with the impacts of drought and other shocks normally will not be able to save enough to augment their endowment of productive assets and increase their potential to generate income.

If building resilience can contribute to poverty reduction, the converse is also true. Reducing poverty can be a way to increase resilience, but reducing poverty does not automatically result in enhanced resilience. Resilience is determined by the three factors described above—exposure, sensitivity, and coping capacity. For the purposes of this book, to allow estimation of the numbers of people who are resilient, the poverty line was used to determine coping capacity: households that see their income fall below the poverty line following a shock were deemed unable to cope (that is, these households were considered non-resilient). Households that following a shock see their income remain above the poverty line were deemed able to cope (that is, they were considered resilient).

Whether a given household will see its income fall below the poverty line following the occurrence of a shock depends on the household's income level before the onset of the shock, its degree of exposure to the shock, and the sensitivity of its livelihood strategy to the effects of the shock. Relatively poor households that started out just above the poverty line may be considered resilient if they are not highly exposed to the shock or if their income is not sensitive to the effects of the shock. Likewise, relatively wealthy households that started out well above the poverty line may be considered non-resilient if they are highly exposed to the shock or if their income is extremely sensitive to the effects of the shock. In summary, poverty influences resilience, but it does not in itself determine resilience, and resilience is an essential component of a strategy to eradicate poverty in a lasting manner.

Estimation of 2010 Vulnerability Profiles

Quantifying the Dimensions of Vulnerability across Livelihood Types

People who are not resilient to shocks are vulnerable to them. Resilience and vulnerability are two sides of the same coin and must be defined in a way that makes the two concepts easily measurable for all types of areas and livelihood zones. In particular, to quantify the dimensions of vulnerability, one must be able to answer questions such as the following: *How many people living in dryland zones in East and West Africa are vulnerable? Who are these people, and what are their livelihood strategies? What types of resources are needed by these people to become resilient? And how are the numbers of vulnerable people likely to evolve over the long run as the population grows and the economy transforms?*

Two different models were used to estimate the number of vulnerable people in the different aridity zones: a livestock/pastoral model for the arid and hyper-arid zones (zones 1 and 2) and a crop model for all other zones (3 through 6) (table 2.2).

The livestock model could be used to calculate vulnerability for aridity zones 4, 5, and 6, but would result in some double counting of people affected by drought. Thus, only the crop model was used to calculate vulnerability in the semi-arid and sub-humid areas (zones 4, 5, and 6).

Independent of the model used, vulnerability was always considered the outcome of three dimensions, as the following subsections explain.

Table 2.2 **Models Used for the Resilience Analysis, by Aridity Zone**

Aridity zone	Vulnerability model
1	Livestock/Pastoral
2	Livestock/Pastoral
3	Livestock/Pastoral
4	Crop
5	Crop
6	Crop

People Exposed to Droughts and Other Shocks

People *exposed* to droughts and other shocks were defined as people living in any type of dryland area (hyper-arid, arid, semi-arid, or dry sub-humid). This included both rural and urban people, since the urban poor are directly affected by increases in food prices during droughts.

Because most population data for African countries are not geo-referenced, it was necessary to spatialize UN population data broken down by urban and rural areas, aridity zones, and livestock systems across Sub-Saharan Africa in 2010 using gridding methods routinely used in the literature. A major source was the dataset developed at the Columbia University Center for International Earth Science Information Network under the Global-Urban Mapping Project (for details, see SEDAC 2015).

The *UN World Population Prospect* data provided information for 35 Sub-Saharan African countries, which the International Food Policy Research Institute (IFPRI) gridded as in the Global Rural-Urban Mapping Project, Version One, disaggregated by administrative area, Aridity Index, and livestock system.

Seven classes of aridity were then combined into four aridity zones (hyper-arid, arid, semi-arid, and dry-sub-humid) and one of non-drylands, following the Food and Agricultural Organization's (FAO) classification above (table 2.3). Different livestock indexes were clustered as pastoral and mixed farming.

People Sensitive to Drought

People *sensitive* to drought were defined as the share of people dependent on agriculture, evaluated on the basis of recent International Monetary Fund (IMF) estimates of the employment shares of agriculture (Fox et al. 2013), and assuming that people below working age depend on agriculture in the same proportion as people above working age. All those dependent on agriculture are assumed to be equally sensitive to droughts and other shocks. This is admittedly a simplification, since the income share derived from agriculture varies across households, but the data needed to assess the income share derived from agriculture were not readily available consistently across countries.

Survey-based evidence[1] (figure 2.2) suggests that in dryland areas, the share of income coming from farming and livestock keeping is at least 60 percent of the total, so this assumption should not bias the analysis excessively.

Three representative livelihood strategies were identified for use in both projecting the likely consequences of the ongoing demographic and socioeconomic

Table 2.3 How the Various Aridity Classes Are Aggregated into Aridity Zones

Aridity class	Aridity Index: lower bound	Aridity Index: higher bound	FAO aridity classification	Aggregate classification	CGIAR aridity classification
0	not defined by AI				
1	0	0.03	A. Hyper-arid	Drylands	—
2	0.03	0.05	A. Hyper-arid	Drylands	SRT2
3	0.05	0.20	B. Arid	Drylands	SRT2
4	0.20	0.35	C. Semi-arid	Drylands	SRT2
5	0.35	0.50	C. Semi-arid	Drylands	SRT3
6	0.50	0.65	D. Dry sub-humid	Drylands	SRT3
7	0.65		E. Non-drylands	Non-drylands	—

Sources: Consultative Group for International Agricultural Research (CGIAR) and Food and Agriculture Organization (FAO).
Note: — = not available.

Figure 2.2 Income Sources in Drylands vs. Non-Drylands, Selected East and West African Countries, 2010

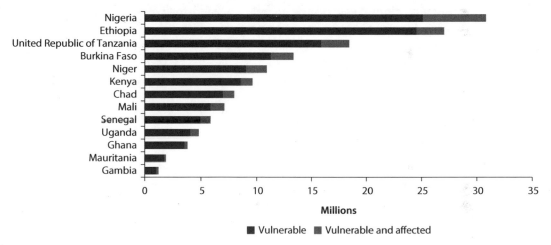

Source: D'Errico and Zezza 2015.

transformation of the drylands and assessing the scope for increasing resilience through technological interventions. The three livelihood strategies are (a) crop production only (farming households), (b) mixed livestock-crop production (agro-pastoralists), and (c) livestock-keeping only (pastoralists). Given the absence of detailed census data, the approach used was to combine information obtained from socioeconomic surveys, mainly those found in the World Bank Survey-based Harmonized Indicators Program (SHIP) database, with estimates from agro-ecological analysis. The resultant livelihood strategies were defined as follows (and summarized in table 2.4):

- *Farming households*—people engaged in crop production only. Their number was estimated from the number of rural households that reported not owning any livestock (in a few countries where data on livestock ownership were not available, expert judgment was used).

Table 2.4 Estimated Agriculture-Dependent Population in East and West Africa, 2010

	Population (millions)	Dependent on agriculture (millions)	Of those dependent on agriculture...		
			Crop farming (millions)	Pastoralism (millions)	Mixed livestock-crop production (millions)
Drylands	247.7	171.2	39.5	26.2	105.4
East Africa	92.2	64.7	17.6	12.7	34.3
West Africa	155.5	106.5	21.9	13.5	71.1
Non-drylands	269.0	195.7	57.3	13.0	125.4
East Africa	109.6	78.2	20.8	4.4	53.1
West Africa	159.4	117.5	36.5	8.6	72.3
Total	516.7	366.9	96.8	39.2	230.8

Source: Population data from UNFPA (United Nations Population Fund); breakdown by aridity zone from the International Food Policy Research Institute.

- *Agro-pastoralist households*—people engaged in livestock production along with farming. Their number was estimated as a residual (that is, those not engaged in crop production only). To calculate the number of people engaged in mixed livestock-crop production, the FAO Gridded Livestock of the World map of livestock production systems was superimposed on the population map. People living in locations associated with mixed crop-livestock systems were assumed to be agro-pastoralists (see de Haan 2016).
- *Pastoralist households*—people engaged in livestock production only. As above, the number of people engaged in livestock production was estimated as a residual of the total population not engaged solely in crop production but relying to some extent on livestock. To calculate the number of people engaged in livestock-keeping only, the FAO Gridded Livestock of the World map was superimposed on the population map. People living in locations associated with livestock-only production systems were assumed to be pastoralists.

In 2010, of the approximately 171 million people living in drylands and dependent on agriculture, about 26 million were pastoralists, 106 million were agro-pastoralists, and 40 million were crop farmers. At the level of individual countries, agro-pastoralists were usually the dominant group, but not always, as the relative importance of the three livelihood strategies varied as a function of local agro-ecological and socioeconomic characteristics.

To estimate the population employed in agriculture, the model used the following formula:

$$P_a = P * \pi_{15_{plus}} * \pi_a$$

where

P_a = Total population (P) multiplied by

$\pi_{15_{plus}}$ = Share of population aged 15 or older and by

π_a = Share of employment in agriculture.

To be consistent with the agricultural employment estimates used later, all individuals over 15 years old (the working-age population) were included. Figures were available at the national level and were assumed to be constant across Aridity Index values. The share of employment in agriculture (π_a) was applied as a constant, national-level coefficient to all aridity classes.

To estimate the population employed in crop farming (P_{ac}), the model used the formula:

$$P_{ac} = P_a * \pi_{ac}$$

where

P_a = agricultural population and

π_{ac} = share of crop farming.

P_a is calculated as indicated above. For countries with SHIP survey data, π_{ac} was calculated as the share of households with the head of household employed in agriculture and engaged exclusively in crop farming (that is, not owning any livestock). For countries without SHIP survey data, proportions of 15 percent for predominantly dry countries and 40 percent for countries along a coast and/or a large lake were assumed. This coefficient was calculated at the national level and applied to all aridity classes as a constant.

To estimate the population employed in pastoralism and mixed farming, the model used this equation:

$$P_1 = (P_a - P_{ac}) * \pi_1;\ P_m = (P_a - P_{ac}) * (1 - \pi_1),$$

where the pastoralist population (P_1) is the population not exclusively engaged in crop farming multiplied by a coefficient measuring the proportion of people living essentially from livestock-rearing (π_1). The mixed-farming population (P_m) is the rest of the non-crop-only population. This last population both grows crops and keeps livestock. P_a and P_{ac} were calculated as indicated above; π_1 was derived by overlaying the gridded population map with FAO's livestock production system map[2] and varies across aridity classes and livestock systems.

People Vulnerable to Droughts and Other Shocks

People *vulnerable* to droughts and other shocks were defined as the proportion of exposed and sensitive people who are also "poor" as derived from the PovCalnet[3] database. Both the crop and the livestock models used the international poverty line of US$1.25 per person per day as the minimum income threshold to be considered resilient to shocks. However, a subtle difference exists between the two, since in the crop model a household was determined to be poor by the fact that it lives below the poverty line, while in the livestock model this depends on the number of tropical livestock units (TLUs) the household must possess to stay at an income of US$1.25 per person per day over the simulation period.

Mitigating Drought Impacts in Drylands • http://dx.doi.org/10.1596/978-1-4648-1226-2

Crop Model

Separate estimates of rural and urban poverty rates are rarely available, so the national (overall) poverty rate was used. The resulting estimates of the number of vulnerable people are undoubtedly conservative, because: (a) poverty is usually higher in rural areas than in urban areas and (b) poverty is usually higher in dryland areas than in non-dryland areas. Recognizing that in drought years people dependent on agriculture experience income losses, in some of the analyses carried out for this book, the number of people unable to cope was estimated using other poverty lines. Based on the sensitivity analysis performed during the design phase of the *Africa RiskView* software,[4] it was assumed that in a year of mild drought, households with incomes exceeding US$1.25 per day by less than 15 percent (US$1.44 per day) would fall below the international poverty line. The same holds for households with incomes exceeding the international poverty line by less than 30 percent (US$1.63 per day) and for those with incomes exceeding the international poverty line by less than 45 percent (US$1.81 per day) in the event of a medium or severe drought, respectively. In each case, the corresponding poverty head count was estimated based on income distribution data obtained from the PovCalnet database.

Using the previous definitions and available data, the different components of vulnerability were estimated for the areas in the drylands of Africa for which the household survey was representative in 2010. Vulnerability profiles define the percentage of the population living in each area likely to be affected by mild, medium, and severe droughts when they occur and the percentage of the population not at any risk from drought.

Throughout the entire region, of the total 424 million people living in drylands in 2010 (exposed to drought and other shocks), approximately 240 million were dependent on agriculture (sensitive to droughts and other shocks). Of this last number, some 97 million people were living below the poverty line and hence vulnerable to droughts.

In East and West Africa, the two subregions of focus in this book, the equivalent numbers were 306 million people exposed, 186 million people sensitive, and 71 million people vulnerable to droughts and other shocks. Most of those exposed to droughts and other shocks were the people living in the driest zones, including the hyper-arid, arid, and semi-arid zones. In these three zones, the population unable to cope with the effects of droughts and other shocks was on the order of 46 million people, or roughly 15 percent of the total drylands populations in East and West Africa (table 2.5).

Livestock Model

Five simulation models were used to estimate the vulnerability to drought of livestock dependent population, including agro-pastoralist and pastoralists.

(a) **Biogenerator model.** Developed by *Action Contre la Faim*, this model uses the Normalized Difference Vegetation Index and Dry Matter Productivity data

Table 2.5 Three Dimensions of Vulnerability in Africa's Drylands, 2010 (Million People)

Region/aridity zone	Exposed	Sensitive	Unable to cope
East Africa	**150.6**	**96.5**	**29.2**
A. Hyper-arid	4.7	2.9	0.5
B. Arid	30.5	18.8	3.9
C. Semi-arid	64.5	41.7	11
D. Dry sub-humid	50.9	33.1	13.8
West Africa	**155.5**	**89.8**	**42.3**
A. Hyper-arid	0.9	0.5	0.2
B. Arid	19.2	12.2	4.8
C. Semi-arid	90.6	53.2	26.3
D. Dry sub-humid	44.8	23.9	11
Subtotal East and West Africa	**306.1**	**186.3**	**71.5**
Central Africa	**13**	**8.6**	**5.1**
B. Arid	0.1	0.1	0
C. Semi-arid	3.2	1.9	0.5
D. Dry sub-humid	9.7	6.6	4.6
Southern Africa	**105.7**	**44.2**	**20.8**
A. Hyper-arid	0.1	0	0
B. Arid	1.8	0.5	0.2
C. Semi-arid	56.8	20.7	7.8
D. Dry sub-humid	47	23	12.8
Grand total	**424.8**	**239.1**	**97.4**

Source: World Bank calculations, PovCalnet database.

collected for the years 1998–2011 from Spot 4 and 5 (Ham and Filliol 2012) on a daily basis in a 1 kilometer by 1 kilometer resolution, aggregated to a 10 kilometer by 10 kilometer grid. The model was used to estimate for each climatic zone spatially referenced usable and accessible biomass (that is, biomass that is edible by livestock). This was done by adjusting total Dry Matter Productivity for the natural vegetation (excluding crop residues) by a factor of 0.3 to account for the poor accessibility of some grassland areas, trampling by livestock, and losses from insects. On the basis of the usable biomass data from 1998 to 2012 and simulated precipitation patterns for past drought and severe drought, the Biogenerator model simulated usable biomass data for the period 2012–30. The mild-drought feed-production projections were based on extrapolating production from years with similar data in the 1998–2011 period. Accessible and usable feed production in severe dry periods was estimated at 90 percent of mild drought yields.

(b) **GLEAM model.** Based on existing databases, the study from the Gridded Livestock of the World (Robinson et al. 2014), georeferenced databases on crop yields and harvested areas (IIASA/FAO 2012), and data from the Biogenerator model, the Global Livestock Environmental Assessment Model (GLEAM) developed by Gerber et al. (2013) calculates at pixel and aggregate level:

- Crop byproducts and usable crop residues
- Livestock rations for the different types of animals and production systems, assuming animal requirements are first met by high-value feed components (crop byproducts if given, and crop residues), and then by natural vegetation
- Feed balances at pixel and aggregate level, assuming no mobility at pixel level and full mobility at grazing shed level
- Greenhouse gas (GHG) emissions intensity

For the drylands under study in this book, the simulation indicated the areas where local feed supplies are adequate on a year-round basis and the areas where feed resources are inadequate and mobility is needed.

(c) **IMPACT model.** On the basis of the feed rations provided by GLEAM, the International Model for Policy Analysis for Agricultural Commodities and Trade (IMPACT) developed by IFPRI was used to (a) calculate the production of meat and milk in drylands and (b) estimate how production will affect overall supply of and demand for these products in the region. At the core of the IMPACT model is a partial equilibrium, multimarket economic model that simulates national and international agricultural markets. It covers 44 commodities, meats, milk, and eggs, specified as a set of 115 country-level supply and demand equations whereby each country model is linked to the rest of the world through trade. For the purposes of this study, the definition of livestock production systems was aligned to match those used by the other models (for example, GLEAM). Based on the changes in supply as a result of the changes in livestock numbers simulated by the MMAGE model (see next) and the variation in feed availability, IMPACT calculated the corresponding changes in per-animal productivity according to the variation in feed availability calculated by GLEAM. The analysis was further informed by the projections of FAO's livestock supply/demand model (Robinson and Pozzi 2011).

(d) **MMAGE model.** This model, based on the R statistical package, consists of a set of functions for simulating the dynamics and production of animals that are categorized by the animals' sex and age class. It was used to calculate the sex/age distribution of the four main ruminant species (cattle, camels, sheep, and goats), their feed requirements in dry matter, and their milk and meat production, data that were in turn fed into the IMPACT and ECO-RUM models (see next).

(e) **ECO-RUM model.** Developed by Agricultural Research for Development (CIRAD) under the umbrella of the African Livestock Platform (ALive), ECO-RUM is an Excel-supported herd dynamics model based on the earlier International Livestock Center for Africa (ILRI)/CIRAD DYNMOD. The ECO-RUM model was used to estimate the socioeconomic effects of changes in the technical parameters of the flock or herd (for example, return on investments, income, and contribution to food security). For the ECO-RUM and MMAGE models, the technical parameters of fertility, mortality, growth, and production of meat, milk, fibers, and organic fertilizer, were obtained based on (a) an FAO/CIRAD/World Bank expert consultation with about 30 specialists, (b) a large literature review carried out by CIRAD, and (c) the authors' experience. In this study, the ECO-RUM model was used to determine the minimum number of TLUs[5] required to provide an income above the poverty level of US$1.25 per person per day for the 2012–30 period, given assumptions on weather, technical parameters, prices, and additional income from outside the sector. This was done interactively for a herd of a standard species composition (for details, see de Haan 2016).

Figure 2.3 provides a schematic overview of the links between the five livestock models.

The results of these models were input into the final step of the analysis: assessing the number of households falling into each of these three categories: (a) resilient

Figure 2.3 Integration of Livestock Models

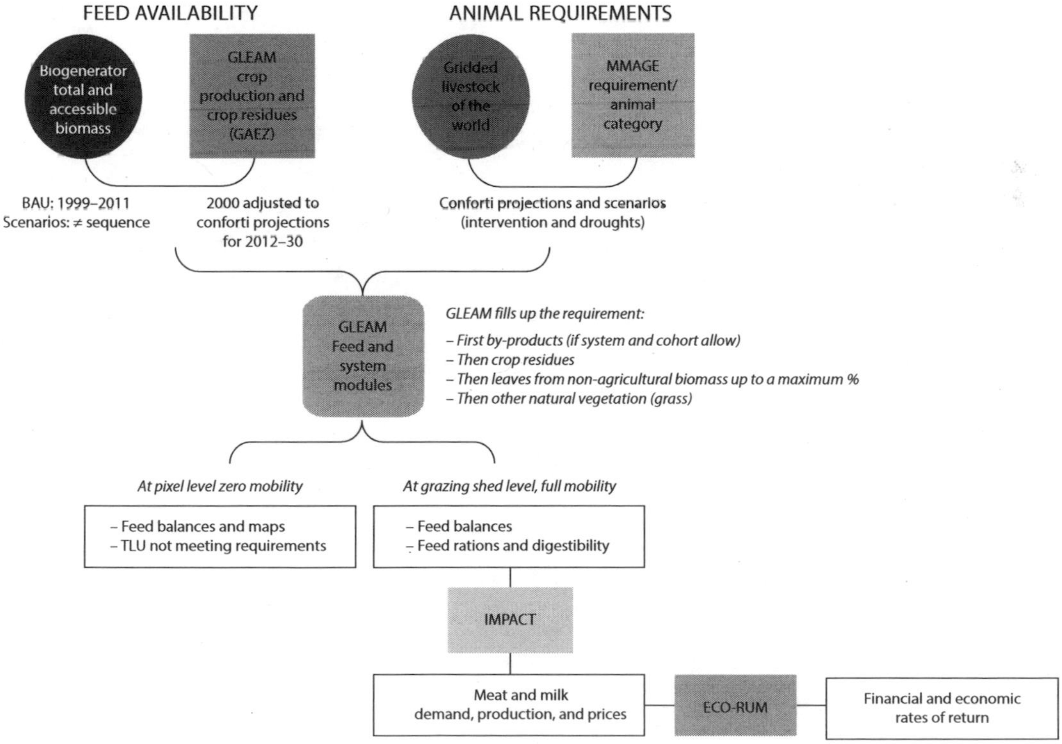

Source: De Haan 2016.
Note: BAU = business as usual; GAEZ = Global Agro-Ecological Zones database: FAO; GLEAM = Global Livestock Environmental Assessment Model; IMPACT = International Model for Policy Analysis for Agricultural Commodities and Trade; TLU = tropical livestock unit.

households, (b) households vulnerable to shocks, and (c) households likely to move out of livestock-based livelihoods. These three groups were estimated based on their ownership of livestock, measured in terms of TLU. The values of the thresholds were used to classify households into one of the three categories were estimated using ECO-RUM, and the corresponding population shares were calculated using a log-normal estimate of the TLU distribution, which approximates quite well the actual TLU distributions emerging from the SHIP database (figure 2.4).

The share of households' P_t estimated to own less than a certain TLU threshold t was calculated as follows:

$$P_t = \int_o^t f(\tau,\mu,\sigma)d\tau$$

where $f(\tau, \mu, \sigma)$ is the log-normal probability distribution function:

$$\sigma = \sqrt{2}\Phi^{-1}\left(\frac{G+I}{2}\right)$$

where $\Phi\text{-}1$ () is the inverse function of the standard cumulative normal distribution; G is the Gini coefficient, calculated from SHIP survey data (table 2.6) as

$$\mu = ln(f) - \frac{\sigma^2}{2};$$

Figure 2.4 Burkina Faso: Cumulative Distribution of Cattle Ownership

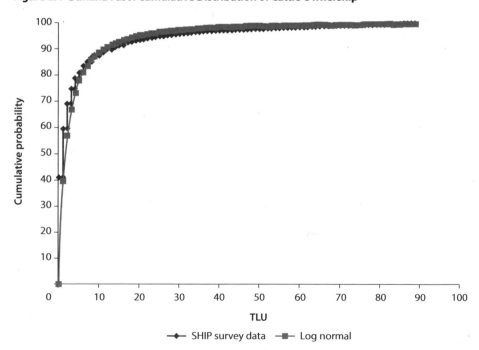

Source: World Bank calculations based on SHIP survey data.
Note: SHIP = Survey-Based Harmonized Indicators Program; TLU = total livestock unit.

Table 2.6 Gini Coefficient of Livestock Ownership, Selected West and East African Countries

Country	Survey year	Income Gini	Livestock Gini	Notes and sources
Burkina Faso	2003	39.60	52.1	Survey did not include medium-size livestock (ovine).
Chad	2011	39.78	74.0	Source: *Troisième Enquête sur la Consommation*
Ethiopia	2011	33.60	55.4	
Kenya	2005	47.68	78.1	Excluded TLU > 2,000 (considered outliers).
Mali	2010	33.02	57.8	Livestock Gini was estimated based on Income Gini.
Mauritania	2008	40.46	66.5	Livestock Gini was estimated based on Income Gini.
Niger	2007	43.89	67.3	
Nigeria	2004	42.93	76.6	Excluded TLU > 1,500 (considered outliers).
Senegal	2005	39.19	76.1	
Tanzania	2007	37.58	67.3	Survey did not include medium-size livestock (ovine); excluded TLU > 5,000 (outliers).
Uganda	2005	42.62	54.7	Calculation only includes medium-size livestock (ovine); figures on large-size livestock appear dubious.

Source: De Haan 2016.
Note: TLU = total livestock unit.

and where *t* is the average number of TLU per household, calculated by dividing the estimate of total TLU for the relevant country/production system by the corresponding estimated number of households.

The inequity of livestock wealth distribution in pastoral societies is striking, and, as discussed below, is one of the major reasons why policy reforms should accompany technological innovation.

In addition to the Gini coefficient (which was assumed constant throughout the simulation, with the exception of parametric reductions used to simulate the effect of redistribution policies), the other key parameter that determines the number of households below or above the thresholds is the average number of TLU/household. This figure, as projected by MMAGE and calculated on the basis of the feed supply estimated by the Biogenerator and GLEAM models, was estimated by dividing the total number of TLU in each dryland country by the total number of households in that country (see "Estimation of 2030 vulnerability profiles"). Details on the TLU and livestock-dependent household estimates by country and livestock production system for 2010 are contained in de Haan (2016).

For this analysis, a pastoral household was assumed to comprise six persons, with 70 percent of its income derived from livestock and a livestock herd consisting of 50 percent cattle, 25 percent sheep, and 25 percent goats. This combination is referred to in the rest of this book as a "pastoral household." Agro-pastoral households were assumed to derive 35 percent of their income from livestock.

The ECO-RUM model estimated that 14.8 TLU was the critical TLU threshold for pastoral households to reach the income of US$1.25 per person per day over 2012–30. This was based on feed production over the 1998–2011 period as determined by the Biogenerator and GLEAM models.

Figure 2.5 Share of the Pastoral Population (%) above the Resilience Level (2010), by Country, Disaggregated by Pure Pastoralists and Agro-Pastoralists

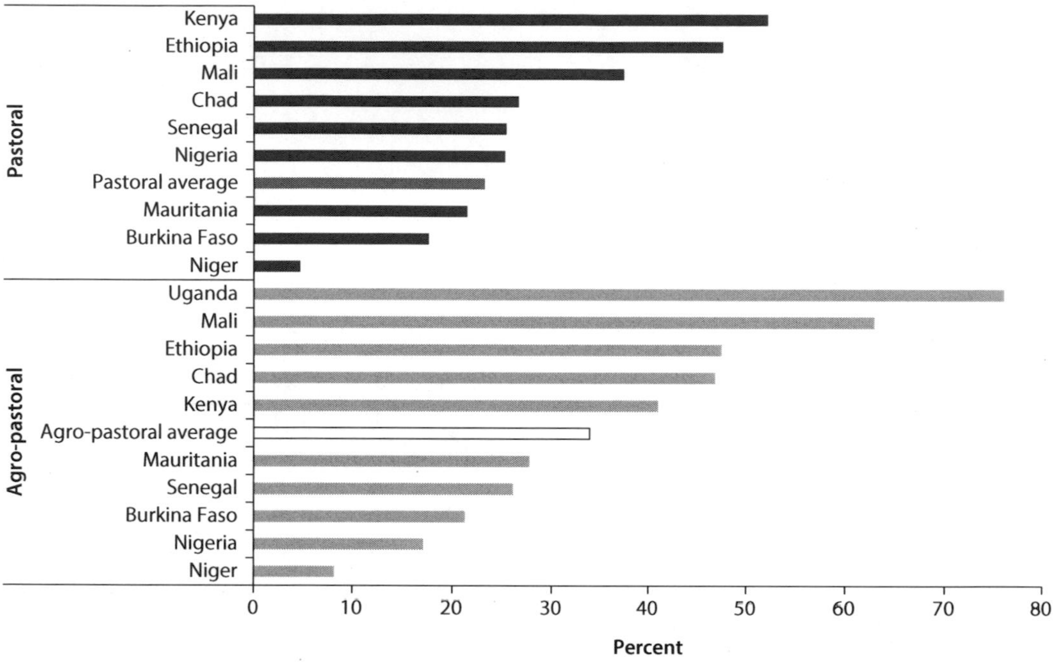

Source: De Haan 2016.

Figure 2.5 illustrate the shares of households calculated to be above the resilience threshold in 2010.

Estimation of 2030 Vulnerability Profiles

The key elements of the model are summarized in this subsection, which also describes the main features of the 2030 "business as usual" (BAU) baseline scenario, in which no interventions are implemented to reduce the number of drought-affected people. To enable comparison with the 2010 baseline figures, the model produced projections for 2030 of the three components of vulnerability, as described in the following.

People Exposed to or Sensitive to Droughts

People exposed to droughts. People *exposed* to droughts were defined as those living in drylands in 2030. The number was obtained by spatializing the UN population projections in accordance with the Global Urban Mapping Project dataset used to determine the 2010 baseline. Differences in urban and rural rates of growth are built into the UN projections, reflecting the ongoing trend toward increasing urbanization. Three sets of estimates were generated, one for each of the three UN fertility scenarios (low, medium, and high). As with the 2010 baseline, the numbers were disaggregated by aridity class and subnational juris-diction for each scenario.

People sensitive to droughts. People *sensitive* to droughts were defined as those living in drylands in 2030 and dependent on agriculture. Because economic growth in dryland countries is likely to be accompanied by structural transformation, the share of agricultural employment in total employment is projected to decline; therefore, the model scaled down agricultural employment as a function of economic growth (the scaling factor was derived from a cross-country regression using a large sample of developing countries worldwide). GDP growth per capita in 2030 was calculated for each dryland country by applying to the 2010 baseline growth an increase estimated on the basis of historical GDP growth recorded in each country during the period 1980–2010. To accommodate uncertainty about future GDP growth, three scenarios were modeled (slow, medium, fast), reflecting the 25th, 50th, and 75th percentiles of the distribution of the historical average growth rates (each average in the sample was calculated based on a 20-year period). The *World Development Report 2008* on agriculture and development showed how globally, starting at US$600 of per capita GDP, a 2.5 percent growth rate over 20 years could lead to a reduction in the labor share of agriculture from 52 percent to 40 percent, thereby possibly reducing relative exposure to shocks (World Bank 2007) (figure 2.6).

Figure 2.6 Share of Agriculture in Total Employment, Selected Lower- and Middle-Income Countries

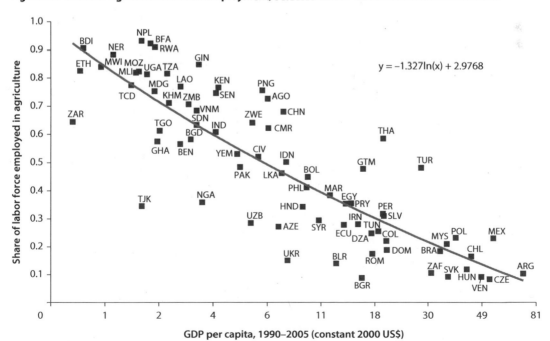

Source: World Bank 2007.

To calculate projected GDP growth rates between 2010 and 2030, past GDP trends were analyzed for each country. The historical data spanned from 1970 to 2012, and observations on the evolution of GDP rates were based on 20-year blocks within that range (1970–89, 1971–90, and so on till 1993–2012) as follows:

- Countries were clustered in three groups to maximize the within-group similarity as given by the coefficient of correlation. Given XLSTAT's agglomerative hierarchical clustering, each cluster was built so the correlation among past GDP growth rates (by 20-year blocks) was highest. The idea was that an underlying process drives GDP growth in each cluster of countries, so it would be wrong to estimate the statistical properties of GDP outcomes assuming independent distributions.
- For each cluster, the statistical distribution that best fit the data in each country was identified (using @RISK), taking into account the correlations among GDP growth rates.[6]
- A Monte Carlo simulation was then run on the distributions defined, to see how the GDP outcomes would be jointly determined, taking into account the correlations among countries. The results of the simulation are with statistical estimates of mean, bottom 25 percent, and top 75 percent of the distribution of GDP rates over 20-year blocks.

The calculated growth rates were then applied over 20 years, and GDP per capita was computed for each of the three 2030 fertility scenarios using their estimated populations.

People Vulnerable to Droughts and Other Shocks

People vulnerable to droughts and other shocks were defined as people living in drylands in 2030, dependent on agriculture and livestock, and living below either (a) the international poverty line (US$1.25 per day) in semi-arid and sub-humid areas or (b) the minimum TLU threshold in arid and hyper-arid areas.

Crop model. The crop model can compute all results using different GDP growth assumptions: bottom 25 percent, mean, and top 25 percent of GDP rate distribution. The number of people living in poverty was calculated by applying to 2030 per capita GDP (estimated as described above) an elasticity coefficient representing the growth elasticity of poverty reduction (GEPR). GEPR measures by how much a 1 percent increase in GDP growth rate could lead to a reduction of the poverty rate.

To estimate the evolution of poverty by 2030, or GEPR, for each country, the first quartile, median, and third quartile of the GEPR were estimated based on historical World Bank data, as follows:

- The change in poverty head count comes from PovCalNet data and refers to actual or interpolated survey data, at the national level, for each of

the following years: 1984, 1987, 1990, 1993, 1996, 1999, 2002, 2005, 2008, and 2010.
- The change in GDP per capita comes from World Development Indicators and refers to the same years.
- The elasticity is the ratio between the changes in poverty head count divided by the change in GDP per capita, measured over each of the 10 three-year intervals.
- A negative elasticity means that growth helps reduce poverty; a positive elasticity means that growth increases poverty.

For the two countries where not enough data were available to compute the GEPR (Gabon and The Gambia), a rate of −0.75 was used. Poverty rates were then projected based on GDP growth, population growth (to compute GDP per capita), and GEPR using the following formula:

$$Poverty\ rate\ 2030\ for\ given\ fertility\ scenario = Poverty\ rate$$
$$2010*[(1\text{-}GEPR)*(GDP\ per\ capita\ 2030 -$$
$$GDP\ per\ capita\ 2010)/GDP\ per\ capita\ 2010]$$

Note that at this stage, the model accounts not only for the working-age population but for the whole population under each category (including children ages 0–14). To accommodate uncertainty regarding the degree to which future growth will result in poverty reduction, three scenarios were modeled: (a) pro-poor growth (GEPR takes on the 75th percentile of the distribution of values observed over the past 20 years), (b) non-pro-poor growth (GEPR takes on the 25th percentile of the distribution observed over the past 20 years; and (c) intermediate case (GEPR fixed at 0.75 for all countries).

This approach was designed to capture the overall experience of growth in Africa, which often has not been particularly pro-poor, while avoiding potential distortions that could result if the most recent GDP growth and GEPR values were simply extrapolated (since both parameters may have experienced short-term upward or downward spikes).

Livestock model. The analysis focuses on the estimation of the evolution over time of the share of households in the following categories: (a) resilient (that is, those more than US$1.25 per person per day over the 2011–30 period), (b) vulnerable (that is, those between US$0.30 and US$1.25 per day per person), and (c) forced or pushed out (those between US$0.01 and US$0.30 per person per day). Given the information described above, the ECO-RUM model defined for each household category (as defined by climate scenario and intervention) the number of TLU corresponding to the above income levels. For example, for the baseline weather scenario without intervention, the ECO-RUM model calculated that 14.8 TLU per household were required to reach the resilience level. A detailed breakdown by climate and intervention is provided in de Haan (2016).

Three climate scenarios were simulated over the 2012–30 period, based on 1998–2011 historical data:

- *Baseline weather*: Extends the 1998–2011 basis to 2012–30
- *Mild drought scenario*: 10 mild drought years of –2.0 to –0.5 standard deviation (SD) of the precipitation, 3 average years with –0.5 to 0.5 SD, and 7 good years with 0.5 to +2.0 SD
- *Severe drought scenario*: 3 years of severe drought with less than –2.0 SD of rainfall, 7 mild drought years of –2.0 to –0.5 SD, 3 average years with –0.5 and 0.5 SD, and 7 good years with 0.5 to +2.0 SD.

The number of TLU that could be maintained in the pastoral zone was calculated based on consumption of 2,300 kilograms (kg) feed per TLU per year, assuming that only 30 percent of the feed is accessible and consumable. This potential number of TLU was compared with the herd projections from MMAGE. The lower of the two numbers was taken as the TLU population over the 2012–30 period. To calculate the number of households, the UN population projections (medium-fertility scenario) were adopted.

On the basis of the total TLU and household numbers, the average number of TLU/household was calculated. Given this data point for the average and the estimated Gini coefficient derived from SHIP survey data (assuming a log-normal distribution of animals per household), the share of households and the TLU population in each of the three categories was calculated. The log-normal distribution needs the estimation of two parameters, σ and μ, respectively the SD and average of the variable under consideration.

Approximate farm gate prices of US$2,400 and US$3,000 for metric tons of beef and mutton, respectively, and US$0.72 for liter of milk were used in calculating each household's income. These meat prices were close to world market prices in 2012; the milk price was based on local survey data. The main cost item for pastoral production systems is veterinary costs, estimated at US$0.50 per TLU in the BAU situation and US$2.00 per TLU in the improved situation.

On this basis, figure 2.7 provides for 2030 the share of the population above the resilience level of US $1,900 per household by country. Detailed results regarding the share of the population corresponding to the three categories described above are provided further in a subsequent section in this chapter (see "Resilience analysis for livestock systems"). The picture that emerges is quite alarming, and a more concerted effort will be needed to avoid massive poverty, food dependency, and increased criminality.

Moving from "Vulnerable to Drought" to "Affected by Drought"

Among the people who are exposed, sensitive, and unable to cope, only some will actually experience a drought or other type of shock in any given year.

While the three drought vulnerability levels (mild, medium, and severe) are constant for households assessed as vulnerable in each of the model's time slices

Figure 2.7 Share of the Pastoral Population (%) above Resilience Level (2030) under the Baseline Scenario, by Country

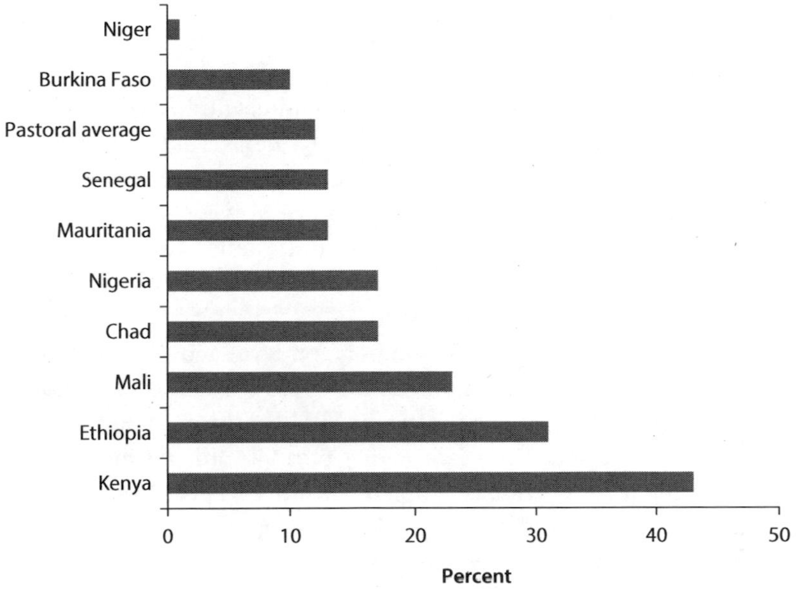

Source: De Haan 2016.

(2010 and 2030), the frequency, geographical scale, and severity of shocks are stochastic and vary considerably from year to year. To estimate the annual number of people affected by drought, the model used the methodology developed by the African Risk Capacity (ARC) team and underlying the Africa RiskView (ARV)[7] software.

The objective of ARV is to estimate the number of people affected by a drought event during a rainfall season and then the dollar amount necessary to respond to these affected people in a timely manner. To do this, ARV translates satellite-based rainfall information into near real-time impacts of drought on agricultural production and grazing using existing operational early warning models. By then overlaying this data with vulnerability information, the software produces a first-order estimate of the drought-affected population and, in turn, response cost estimates.

In the original ARV model satellite rainfall data are converted into a drought index, known as the Water Requirement Satisfaction Index (WRSI), as an indicator of crop performance based on the availability of water to a crop during a growing season. The WRSI captures the impact of timing, amount, and distribution of rainfall on staple, annual, rainfed crops.

In ARV, the WRSI drought index can range from 0 to 100, where 100 indicates no water deficit for a crop and therefore no expected water deficit-related reduction in yield. A number less than 100 indicates some water deficit stress and therefore some expected yield reduction as a result.

Mitigating Drought Impacts in Drylands · http://dx.doi.org/10.1596/978-1-4648-1226-2

For this analysis, the Africa RiskView software methodology was adapted substituting the WRSI with simulations of the IFPRI's DSSAT[8] Cropping System Model v4.5 as an input.

For the purposes of identifying a drought and then estimating populations affected, the drought index is aggregated at the level of a "vulnerability polygon." Vulnerability polygons are specific geographical units within a country for which information on household vulnerability to drought exists and is statistically representative; they are usually administrative units or livelihood zones. The drought index is aggregated at the vulnerability polygon level by taking the arithmetic mean of all drought index grid point values that fall within each polygon. In this study, a polygon results from the intersection of each country's first administrative level and the aridity zone.

Drought in ARV is determined at the polygon level and is defined as a negative deviation of the aggregated drought index at the end of a rainfall season from its normal level. In its default settings, ARV considers the median drought index value of the previous five years as the "normal" drought index for a polygon. To determine if a drought has happened at the end of a season, the drought index for the season and polygon in question is compared against the normal for that polygon. If the drought index value for the season is between 81 percent and 90 percent of the normal, ARV considers there to have been a mild drought. If the value is between 71 percent and 80 percent of the normal, ARV considers there to have been a medium drought. A severe drought is defined as any drought index value that is less than 71 percent of the normal. The value at which drought severity begins is referred to as a "drought trigger," summarized as follows:

- *Mild* drought trigger: 90 percent of normal, 10 percent deviation
- *Medium* drought trigger: 80 percent of normal, 20 percent deviation
- *Severe* drought trigger: 70 percent of normal, 30 percent deviation.

Once the drought index is compared to its benchmark and the severity of a drought in a polygon is defined, a scaling factor is applied to translate the drought index deviations from normality into agricultural production losses and in turn into agricultural income loss. The default value for this parameter is currently 1.5, derived from a relationship of 1:1.5 between drought index percentage deviation from the benchmark and agricultural production loss, and a relationship of 1:1 between the latter and a decrease in agricultural income.

Using the simulations of the IFPRI's DSSAT Cropping System Model v4.5 (Jones et al. 2003; Hoogenboom et al. 2010) as an input instead of the drought index, the ARV model can simulate the impact of drought with and without the best-bet technologies. To avoid potential distortions associated with using yield estimates instead of drought index values, it was assumed that the differences in crop yields attributable to adoption of the best-bet technologies translate into equivalent differences in agricultural income (in the ARV model, this is tantamount to setting the scaling factor to a value of 1, as using yield data as an input does not require the model to assume the 1:1.5 relationship between drought index deviation and agricultural production loss).

The next step in the ARV methodology was to convert this information into a drought-affected population estimate for that polygon. To do this, each polygon was given a household drought vulnerability profile.

To adjust the model for the use of DSSAT-based input data, the threshold deviations from drought index that define mild, medium, and severe drought based on the 1:1.5 relationship between drought index deviation and income losses were therefore scaled to the 1:1 relationship between yield deviation from normality and income losses. These profiles are static within ARV and are calculated, as explained above, outside of the software from available household survey data such as government and World Food Programme Comprehensive Food Security and Vulnerability Analysis surveys, which are available in many countries.

Once the vulnerability profiles were set, their use in ARV to estimate affected populations was straightforward, according to the following formula:

$$\text{If } y > TMl^*Y \text{ then } N = 0$$

$$\text{If } TMl^*Y \geq y > TMd^*Y \text{ then } N = Ml^*P + b1 * (TMl^*Y - y)$$

$$\text{If } TMd^*Y \geq y > TS^*Y \text{ then } N = Md^*P + b2 * (TMd^*Y - y)$$

$$\text{If } y \leq TS^*Y \text{ then } N = S^*P$$

$$b1 = (Md^*P - Ml^*P)/(TMl^*Y - TMd^*Y), \text{ and}$$

$$b2 = (S^*P - Md^*P)/(TMd^*Y - TS^*Y),$$

where P is the population of the polygon with a vulnerability profile of

- Ml percentage of the population at risk to mild drought
- Md percentage of the population at risk to medium drought
- S percentage of the population at risk to severe drought

where N is the estimated number of people affected by a drought in the polygon that has experienced a rainfall season described by y, the seasonal yield value; Y is the benchmark, that is, the "normal" yield for the polygon, usually an average of past years' yields; and TMl, TMd, and TS are the mild, medium, and severe drought triggers, respectively, defined as percentage deviations of the seasonal yield from its benchmark.

In other words, if a mild, medium, or severe drought occurs, defined as the ratio between yield and the benchmark precisely meeting the mild, medium, or severe drought triggers, the people at risk to mild, medium, and severe drought in that polygon are assumed to be affected. All affected population estimates for yield values between the drought triggers for the different severities of drought can be found through a simple process of linear interpolation between the affected population estimates at the drought trigger points (see figure 2.8). If there is no drought, as measured by yields being above the first drought-detection point (10 percent in the example), the estimated population affected is 0; for all yields/benchmark ratios less than the severe drought trigger, the estimated affected population stays the same, since all possible at-risk households in the polygon are assumed to have been affected. A 20 percent deviation below the normal condition of drought index corresponds to 12,000 people affected in a hypothetical polygon.

The specific vulnerability profiles at Admin 1 level (the first level of subnational jurisdiction) created for 2010 and 2030 were calculated as explained in the previous subsections ("Estimation of 2010 vulnerability profiles" and "Estimation of 2030 vulnerability profiles"). The 2030 profiles were based on a number of assumptions about demographic increases, economic growth, and structural transformation (as described previously under "Moving from 'vulnerable to drought' to 'affected by drought'") that determine how the number of people below the poverty line and the percentage of people employed in agriculture will change by 2030. Within each Admin 1 level unit, the vulnerability profiles can be broken down further by aridity zone. Vulnerability profiles for 2010 and 2030 under the medium-fertility scenario are available for the majority of East and West African countries. Table 2.10 shows the vulnerability profiles for 2010 and 2030 for the three drought cases in Mauritania. For illustrative purposes only, the crop model-based vulnerability profile was used.

Figure 2.8 Estimated Population Affected in a Polygon as a Function of Deviations in the Drought Index from the Benchmark

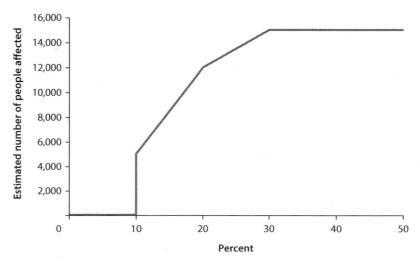

Source: African Risk Capacity 2015.

Table 2.7 Mauritania Vulnerability Profile (Population, Millions)

Region	Aridity class	Mild drought		Moderate drought		Severe drought	
		2010	2030	2010	2030	2010	2030
Assaba	Arid	0.101	0.141	0.122	0.17	0.14	0.196
Brakna	Arid	0.094	0.132	0.113	0.159	0.131	0.183
Gorgol	Arid	0.095	0.134	0.115	0.161	0.133	0.186
Gorgol	Dry semi-arid	0.001	0.001	0.001	0.001	0.001	0.001
Guidimaka	Arid	0.031	0.044	0.038	0.053	0.044	0.061
Guidimaka	Dry semi-arid	0.043	0.06	0.052	0.073	0.06	0.084
Hodh Ech Chargui	Arid	0.115	0.161	0.139	0.195	0.16	0.224
Hodh El Gharbi	Arid	0.087	0.123	0.106	0.148	0.122	0.171
Tagant	Arid	0.021	0.029	0.025	0.035	0.029	0.041
Trarza	Arid	0.092	0.129	0.111	0.155	0.128	0.179
Total		0.68	0.953	0.821	1.15	0.947	1.327

Source: Calculations based on data from African Risk Capacity Agency 2015.

Table 2.8 Coverage of Resilience Interventions

Livelihood	Intervention	Dryland type	
		Hyper-arid, arid	Semi-arid, dry sub-humid
Livestock-based	Livestock-based Improved animal health services	☑	
	Early offtake of young male animals	☑	
	Increased feed supply	☑	
	Equitable access to land and water	☑	
	Increased taxation of wealthy households	☑	
Farming-based and mixed	Drought-tolerant germplasm		☑
	Heat-tolerant germplasm		☑
	Soil fertility management		☑
	Agroforestry/Farmer Managed Natural Regeneration (FMNR)		☑
	Heat-tolerant germplasm and FMNR		☑
	Drought-tolerant germplasm and soil fertility management		☑
	Drought-tolerant and heat-tolerant germplasm		☑
Irrigation			☑

Source: World Bank data.

The definition of mild, medium, and severe drought was kept the same in both the 2010 and 2030 profiles. Furthermore, since the poverty line of US$1.25/day was used in both the 2010 and 2030 vulnerability profile definitions, a comparison of these two baseline profiles (BAU) indicates how economic growth and structural transformation are likely to affect the proportion of the population vulnerable to drought as defined by the ARV model. For example, in Mauritania, even though the share of the total population that is poor is projected to decline, the absolute number of people vulnerable to drought will actually increase by some 40 percent.

Combining the information obtained from the vulnerability profile of figure 2.8 (showing how many people are affected by a certain deviation of yields from the benchmark) with statistical information on the occurrence of drought episodes, one can calculate the number of people affected by drought per year on average. It is important to note that the vulnerability curve shown in figure 2.8 is not linked to the frequency nor to the risk of drought occurring in a particular polygon. For this reason, adoption of a given crop-farming technology does not change the vulnerability profile prevailing in that polygon. Rather, the changes in the mean level and distribution of crop yields registered in that unit following adoption of the technology affect both the seasonal yield and its benchmark, and therefore affect the probability of hitting the drought-specific threshold.

To capture the effect in 2030 of adopting a new or an improved technology, it is necessary to maintain the definition of drought in the model (in terms of the benchmark and drought triggers) and then to calculate the changes in expected number of people affected by drought for all likely yield projections for the various types of technology implemented and for the BAU scenario. For example, consider first the non-intervention scenarios and the medium-fertility scenario. Assume that the rainfall and resulting crop yields that can occur in an area in 2010 and 2030 come from the same distribution; that is, no change in climate occurs for the next 25 years. The DSSAT model generates a series of yields for 25 years for each Admin 1 level/aridity zone unit under the assumption that no technology is implemented in the cropping areas. These same 25 yield values can be imposed on the 2010 and 2030 vulnerability profiles to estimate possible drought-affected populations in those scenarios. Figure 2.9 shows the estimated number of drought-affected people in Mauritania using the 25 yield values.

To estimate the impacts of crop-farming technologies on vulnerable populations, the DSSAT model simulated how the various technologies affect the mean level and distribution of yields. Distributions of the drought-affected population estimated using the yield values from the 25 simulation years for each technology can be compared to distributions of drought-affected populations estimated under the baseline scenario in which yields do not benefit from adoption of any of the technologies.

The differences in estimated numbers of people affected show the impact of each technology on the drought-hit population or, in other words, on households' resilience. Figure 2.9 shows, again for the case of Mauritania, the effects of adopting one of the interventions considered in the analysis (specifically, the adoption of a crop variety that is both drought-tolerant and heat-tolerant). As can be seen, the difference between the 2010 and 2030 no-intervention (BAU) scenarios is given only by the difference in vulnerability profiles; hence it is proportional to the level of drought experienced in any given year. Instead, compared to the 2030 BAU scenario, the number of drought-affected people declines in many years; in some years, the result is only to slow down the increase in the number of drought-affected people, while in other years the number of drought-affected

Figure 2.9 Africa RiskView Estimates of Drought-Affected People in Mauritania Expected for Each of 25 Simulated Yield Years

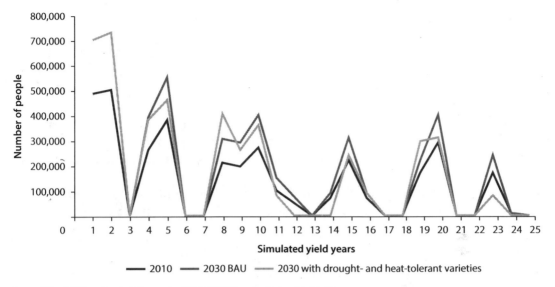

Source: Africa Risk Capacity calculations, using IFPRI's DSSAT Cropping System Model v4.5.
Note: ARV = Africa RiskView; BAU = business as usual.

people actually falls below the 2010 baseline. Overall, adopting the drought- and heat-tolerant variety leads to an 11 percent decrease in the number of drought-affected people. This example shows the benefit of a single intervention adopted in all polygons where, although it cannot prevent the drought from occurring, it can definitely reduce its impact on the population living in the area.

The model assumes that a technology is considered for adoption in a given area if and only if the number of people affected after its implementation is actually lower than in the case of non-intervention. Benefits are maximized when the entire set of interventions is considered, and in each area the technology adopted is the one that leads to the largest reduction in the number of drought-affected people. The results presented in this book are based on the latter approach.

To estimate the number of people affected and to evaluate the impact of adoption of each crop technology, the livestock model described above was used to create a vulnerability profile for aridity zones 1, 2, and 3. The number of people affected in these areas was calculated using the percentage of drought-affected people out of the total vulnerable population observed in the crop model for the same admin unit or a neighboring one.

Irrigation is considered to be an intervention that saves people from being affected by drought. Thus the irrigation model output (see "Resilience analysis for irrigation") can be directly considered as the benefit of implementing irrigation interventions, and the number of people "saved" can be compared to the reduction in people affected because of adoption of a technology or livestock intervention.

Interventions

For the livestock sector, in addition to the technological interventions of improving health and introducing early offtake of male animals for fattening in higher rainfall areas, the background paper simulates: (a) an additional technology option of increasing feed by safeguarding the mobility of herds, promoting feed (hay) markets, and opening unused areas through water development (in arid regions, mainly), which is assumed to be able to increase by 2030 the percentage of accessible feed from 30 percent to 45 percent and (b) policy options addressing the strong inequity in the livestock holdings, by (1) enhancing equitable access to water and grazing through modifications in land use, exclusively allocating user rights to vulnerable households by allocating 50 percent of grazing area exclusively to (groups of) vulnerable households and (2) introducing higher tax rates for wealthy households by lowering the Gini coefficient by 25 percent from the 2010 level.

Resilience Analysis for Livestock Systems

Lower TLU thresholds imply that for a given distribution of livestock as sets and climate (thus, feed availability), more households will be above the threshold and fewer households below, compared to the BAU scenario. Interventions in the improvement of animal health services reduce the mortality rate and increase the number of animals that can be sold, thereby either reducing the number of TLU needed to reach a certain level of income or achieving a better income with the same herd. Similarly, interventions that promote the sale of animals at a younger age for fattening in higher rainfall areas increase the price received per animal and reduce overall mortality, again reducing the number of TLU needed to reach a certain income level.

Given the log-normal distribution shown in figure 2.4 and the vulnerability profiling for the livestock model described earlier, the impacts of climate and intervention scenarios were estimated.

Resilience Analysis for Rainfed Cropping Systems

As with the case of livestock, the model estimated the potential impacts on resilience of interventions targeting rainfed cropping systems. The analysis was conducted in two stages. In the first stage the objective was to estimate the potential impact of adoption of a given crop-farming technology on the yields of crops grown by agro-pastoralist and crop-farming households. In the second stage the objective was to estimate how these yield changes likely translate into income changes and how these in turn affect agro-pastoralist and crop-farming households' resilience to drought. Results are expressed as the number of people affected by drought under the adoption of different technologies.

Modeling Technology Impacts on Crop Yields

The potential impact of the adoption of crop-farming technologies on the yields of crops grown by agro-pastoralist and crop-farming households was estimated using IFPRI's grid-based crop modeling platform. Because it is impractical to model the full range of crops grown in the drylands, the analysis was carried out using the dominant cereal crop grown in any given location, identified with the help of IFPRI's Spatial Production Allocation Model 2005 (You et al. 2014) in 2,294 grid cells distributed across 16 countries. The dominant rainfed crops are millet and sorghum in arid and dry semi-arid zones, and maize in wet zones, some semi-arid zones, and dry sub-humid zones. The crop yield simulations were conducted using three crop models that are part of the DSSAT Cropping System Model:

- CERES-Maize
- CERES-Sorghum
- CERES-Millet.

The main crop cultivated by country and aridity is shown in map 2.2. Seasonal yields were simulated at the level of each grid cell over a 25-year period. Given

Map 2.2 Crop by Aridity Zone/Admin Unit

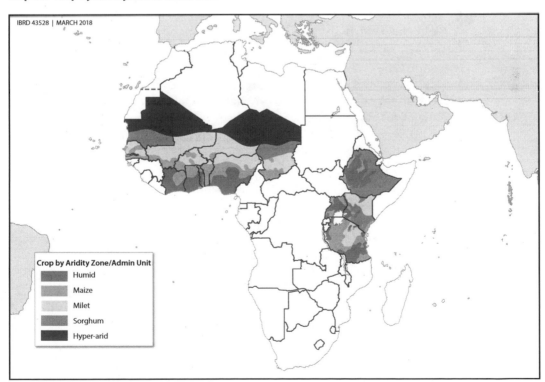

Source: African Risk Capacity Agency 2015.

the assumption that the distribution of rains will remain stable over time, that is, that rainfall patterns in the drylands during the next 25 years will not be significantly different from weather experienced during the past 25 years, daily weather data from 1984 to 2008 were input (Ruane, Goldberg, and Chryssanthacopoulos 2015). Soil properties in each grid cell were represented using IFPRI's HC27 Generic Soil Profiles Database (Koo and Dimes 2013).

Planting windows for the three representative crops were synchronized with the cropping calendar of the ARV model (table 2.9) by selecting the first month of the corresponding crop and country. Since the model runs on a daily time-step, a specific planting date within the planting month was selected, based on predefined planting conditions on moisture content (more than 40 percent of field capacity) and temperature (between 10°C and 40°C) of the top 30 centimeters of soil depth.

The DSSAT framework was used to assess the potential impact on yields likely to result from the adoption of five best-bet crop-farming technologies: (a) drought-tolerant varieties, (b) heat-tolerant varieties, (c) additional fertilizer, (d) agroforestry practices, and (e) water harvesting in the form of *Faidherbia albida* grown in cereal fields. The potential impact on yields was modeled separately for each technology, as well as for several combinations of technologies expected to have synergies (such as varieties with drought tolerance and heat tolerance, drought- or heat-tolerant varieties grown with additional fertilizer, and drought- or heat-tolerant varieties grown in combination with agroforestry).

Crop-only systems. The following summarizes the assessed yield-impact on crop-only systems of three of the four crop-farming technologies mentioned previously, as well as the impact of water harvesting.

• **Drought-tolerant varieties.** To simulate the likely impacts of adoption of drought-tolerant varieties, which are known to have superior rooting ability in the presence of low levels of soil moisture, the soil profile was adjusted by lowering the lower limit of soil water holding capacity (or permanent wilting point) and increasing the relative soil root growth distribution factor (between 0 and 1) parameters in each soil layer.[9,10] This change effectively enhanced the water extraction capability of crop roots. In the case of maize, the anthesis-silking interval sensitivity was additionally introduced (Rosegrant et al. 2014; Lizaso et al. 2015). This change enabled the CERES-Maize model to simulate the varietal traits of maize to alter the timing of silk emergence as a response to physical stresses. Under drought conditions, typical maize varieties delay the emergence of silks, with respect to pollen shed, resulting in reduced pollinated female flowers and, therefore, decreased grain yield. In contrast, the timing of silk emergence for drought-tolerant maize is less sensitive to water stress, avoiding the potential yield reduction. This mechanism was not available in the original version of the CERES-Maize model, but was implemented by incorporating two new cultivar-specific parameters required to control for the non-stress

Table 2.9 Maize, Millet, and Sorghum Cropping Calendar, Selected West and East African Countries

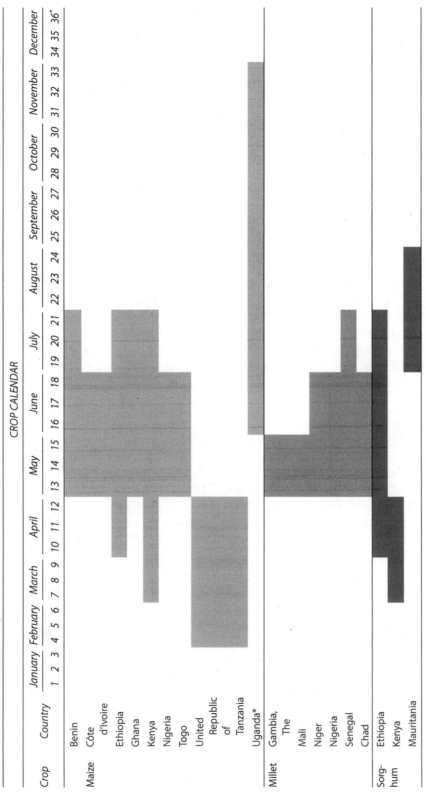

CROP CALENDAR

Source: IFPRI's DSSAT Cropping System Model v4.5; * = based on multiple seasons; ** = dekads.

anthesis-silking interval response and the genotype sensitivity to drought. Other types of potential mechanisms and breeding strategies are used in practice but were not implemented in this study; their benefits to crops were assumed to be reasonably similar to those associated with having more access to water.

- **Heat-tolerant varieties.** Plant growth and yield are highly sensitive to changes in temperature. Each plant species has its own minimum, optimum, and maximum temperatures that define the climatological suitability where it can survive and thrive. These temperature values, collectively called "cardinal" temperatures, are well-documented across plant species (for example, Ecocrop Database [FAO 2013]). In DSSAT, the cardinal temperature parameters are defined in the species parameter files that are shared across all varieties. In the case of maize, for example, the CERES-Maize model defines the optimum and maximum temperatures for grain filling as 27°C and 35°C, respectively, in the species parameter file (MZCER045.SPE). Compared to the baseline heat-susceptible varieties, it was assumed that heat-tolerant varieties would have higher optimum and maximum temperatures by 2°C to allow them to sustain normal growth and grain-filling rates at increased temperatures. To implement this change in cardinal temperatures as the varietal difference within the same species, a separate copy of the species file was created and used in the simulation of heat-tolerant varieties. In this file, all other parameter values were identical to the original file except that the optimum and maximum temperatures increased by 2°C (for example, to 29°C and 37°C, respectively).

- **Additional fertilizer.** The baseline, no-intervention scenario includes an inorganic nitrogen fertilizer application rate that is specific to each region, input system, and crop, and ranges between 10 and 20 kg[N]/ha/season. This was obtained by calibration of simulated raw yields to FAOSTAT-reported, country-level yields (Rosegrant et al. 2014). For the optimal fertilizer intervention, the baseline fertilizer application rate was increased by 50 percent.

- **Water harvesting.** To simulate the potential effects of harvesting runoff and storing it *in situ* for use in supplementary irrigation, a two-stage approach was implemented in the DSSAT modeling framework (Rosegrant et al. 2014). The model was first run without any water management practices, and the model output files, especially the seasonal summary output (file name: Summary.out), the soil water output (file name: SoilWat.out), and the plant growth output (file name: PlantGro.out), were analyzed to identify periods during the growing season when yields are constrained by lack of water. These periods represent opportunities for implementing improved water harvesting and supplementary irrigation practices. The simulation results were also used to determine when supplementary irrigation can have the largest impact on yields (such as immediately after germination and before flowering), and also to estimate how much of the harvested water, assumed collected from the runoff water, would be available from *in situ* storage. The model was then run again including harvested runoff water in the form of supplementary irrigation, applied during the identified irrigation window.

Agroforestry

Faidherbia drops its leaves just prior to the rainy season, which means that in theory the leaves and nitrogen contained in them can be available to crops during the growing season. Many factors affect the availability of the nutrients to the crops, including tillage methods and early season rainfall patterns. However, studies of the effects of *Faidherbia* on yields in farmers' fields have consistently found positive and significant results. To simulate the improvements in soil fertility expected to result from decomposing leaves from *Faidherbia* trees regenerated or planted in the same field as the indicator crops, for each cropping cycle an additional input of organic soil amendments was implemented 10 days before planting. The trees were assumed to be 20 years old in year 1, so that the amount of organic matter contributed throughout the simulation period remains constant. Each tree was assumed to produce 100 kg of leaves, of which 4.3 percent is nitrogen. These values were taken from scientific studies in West Africa (Spevacek 2011).

Two tree density values were simulated (5 trees/ha and 10 trees/ha) to test the sensitivity of crop yields to tree density. Canopy coverage, which determines the area within each field that actually benefits from the decomposition of tree-contributed organic matter, was assumed to be 10 percent and 20 percent for tree densities of 5 trees/ha and 10 trees/ha, respectively. This implies an average tree canopy diameter of 16 meters, which is above the mean found across all landscapes and reflects a healthy tree that was not adversely affected by early browsing or heavy pruning. Although the canopy cover assumption is favorable, the tree density numbers assumed are well within numbers observed on many farms and below the averages in some locations in the semi-arid drylands where FMNR is practiced. Reducing the canopy mean but raising the density (that is, assuming effective extension) will lead to approximately the same effects on crop yields.

Using *Faidherbia* as the sole agroforestry intervention had certain implications for the results. First, it is useful to recall that while *Faidherbia* is distributed throughout the drylands of Africa, it will not emerge through regeneration in all locations. However, it is not possible to model the complex mix of species that will vary over location through regeneration practices in drylands. *Faidherbia* is known to be superior among all species in terms of soil fertility and yield effects. Thus, if *Faidherbia* is replaced by other species, there will be a reduced effect on nitrogen generated and yields. On the other hand, not all of the benefits from *Faidherbia* were valued, including the significant effects on livestock, which feed on its pods. Moreover, by disregarding other valuable species that are likely to emerge in drylands (such as shea), the analysis excluded many other poverty-reducing benefits derived through agroforestry. Whether the model under- or overestimated actual benefits depends on the context (for example, the importance of livestock and tree products to income).

BAU Assumptions

In addition to the representation of simulated technologies described above, assumptions about farmers' baseline (BAU) management practices were made, as follows:

- *Conventional tillage*: Ten days before the planting window starts, conventional tillage is practiced by simulating full soil disturbance in the top 20 cm of soil.
- *Variety selection*: Location-specific variety was identified using the same method described in Rosegrant et al. (2014). This method does not necessarily represent the most commonly found variety found in the area; rather the benchmarking variety was selected to represent the relative yield changes with and without the technology.
- *No manure application*: Other than the organic amendment from agroforestry, no manure application was simulated.
- *No CO_2 fertilization*: Default value of atmospheric CO_2 concentration value of 379 ppm (baseline value of 2005) was used.
- *Harvest scheduling*: Crops are harvested when the plants reach maturity.
- *Full removal of crop residue*: After harvest, all aboveground crop residues are assumed to be removed from fields.

Additional details on the modelling platform setup are available in Rosegrant et al. (2014).

Resilience Analysis for Irrigation

The final intervention evaluated was irrigation development. The assessment built on projected irrigation investment potential in Sub-Saharan African countries. Currently, crop farming in Sub-Saharan Africa is predominantly rainfed; irrigation accounts for only 3 percent of cultivated area in dryland areas, much below the global average. However, enhancing the agriculture sector's performance in Sub-Saharan Africa through irrigation investment is of interest to policy makers.

The methodology for projecting irrigation investment is described in box 2.1. The analysis encompasses both large-scale irrigation (LSI) and small-scale irrigation (SSI) schemes under combined biophysical and socioeconomic constraints. Possible changes in farming area and occurrence of dry-season crop farming induced by irrigation development were modeled explicitly. The constraining factors taken into account include (a) environmental suitability (terrain, accessibility to surface water/groundwater, distance to markets, and so forth), (b) economic viability, (c) water balance/scarcity, and (d) increased demand for irrigated crops related to projected population and income growth.

It is important to note that the cost-benefit and water balance figures used to make the projections are long-term averages and the reported areas with irrigation investment potential are "physical area equipped with irrigation infrastructure." In drought years, when water becomes scarce, irrigation cannot be delivered everywhere, leaving part of the area equipped with unused irrigation infrastructure.

Box 2.1 Projecting Irrigation Expansion Potential in Sub-Saharan Africa by 2030

In the assessment on irrigation investment potential in Sub-Saharan Africa, a distinction was made between large-scale irrigation (LSI) and small-scale irrigation (SSI) schemes. To support the LSI assessment activities, an inventory database containing information (location and capacity) on 680 large reservoirs in Africa was compiled; 120 dams with capacity over 50 MCM (million cubic meters) and at the planning stage or under rehabilitation were included in the analysis. The possible extent of command areas of these reservoirs was delineated via geographic information system (GIS) topographic analysis by assuming that irrigation occurs by gravity and the irrigation area can extend up to 200 kilometers downstream of the dam. For SSI, a specific underlying SSI technology was not assumed. Instead the model was parameterized to approximate situations in which the adoption of several main smallholder irrigation technologies (such as pumps and small reservoirs) could be accommodated.

The projection is performed on the 5 arc minute SPAM (Spatial Production Allocation Model) grid and begins with a GIS mapping analysis to score the environmental suitability of irrigation development of each grid pixel.

Table B2.1.1 Current Inventory of Large Dams in African Countries

	Number of reservoirs	
Dam status	Capacity threshold > 50 MCM (#)	All
Operational	253	489
Planned	106	159
Rehabilitated	14	32
Total	373	680

Source: Xie et al. 2015, based on Africa Infrastructure Country Diagnostic dams database.
Note: MCM = Million cubic meters.

Modeling the variation of "actual" irrigated area becomes important in the context of vulnerability and resilience analysis, and such modeling work using physically based hydrologic and water infrastructure operation models requires detailed data on African aquifers and reservoir operations, which are currently not available. Because of these limitations, a conceptual approach was developed to estimate the impacts of irrigation on drought-affected people in the face of weather variability. The key steps and assumptions used in this irrigation-related resilience/vulnerability analysis are shown subsequently.

Under many circumstances, groundwater provides buffers again drought. Groundwater accessibility was included as a criterion to determine the environmental suitability of SSI development (box 2.1). In the vulnerability/resilience analysis, the SSI areas with good groundwater accessibility and abundant groundwater storage were further delineated using groundwater depth and groundwater storage maps in Africa (table 2.10). It was also assumed that operation of SSI infrastructure in areas with groundwater depth below 25 meters and storage greater than 10,000 millimeters (mm) is free from weather variability.

Mitigating Drought Impacts in Drylands • http://dx.doi.org/10.1596/978-1-4648-1226-2

Table 2.10 Aquifer Classification in British Geological Survey Groundwater Data

Aquifer class	1	2	3	4	5	6
Depth to groundwater (m)	0–7	7–25	25–50	50–100	100–250	> 250
Groundwater storage (mm)	0	<1,000	1,000–10,000	10,000–25,000	25,000–50,000	> 50,000

Source: MacDonald et al. 2012.
Note: m = meters; mm = millimeters.

Table 2.11 Criteria Used to Assess Environmental Suitability of Large-Scale Irrigation Investment (within the Delineated Command Areas of Reservoirs)

Criteria	Range of parameter	Range of score
Topography (slope)	0–10 %	S_1: 100–0
Distance to main channel of river downstream of the dam	0–5 km	S_2: 100–0
Distance to existing irrigation	0–10 km	S_3: 100–0

Source: Xie et al. 2015.
Note: km = kilometers.

Table 2.12 Criteria Used to Assess Environmental Suitability of Small-Scale Irrigation Investment

Criteria	Range of parameter	Range of score
Topography	0–10 % slope	S_1: 100–0
Distance to surface water	0–5 km	S_2: 100–0
Groundwater depth	0–250 m	S_3: 100–0
Travel time to market (hour)	0–3 hours	S_4: 100–0
Distance to existing irrigation	0–10 km	S_5: 100–0

Source: Xie et al. 2015.
Note: km = kilometers; m = meters.

The remaining SSI areas and the LSI areas in operation are affected by drought and might contract in drought years. The modeling of the variation of these areas was based on the same drought characterization method used to analyze the impacts of interventions in rainfed cropping systems. For each ARV vulnerability polygon, the irrigation area in operation was approximated using an exponential function of the chosen drought index:

$$A_i = A_0 * e^{-\alpha I}$$

where A_i is an SSI or LSI area in operation in year i, A_0 is projected area equipped with SSI or LSI infrastructure in 2030 (for SSI, delineated areas with abundant groundwater resources were excluded in A_i and A_0); α (≥ 0) is a constant; and I (0–1) is the drought index. The value of α controls the contraction rate of irrigation area under drought. Reduction in operated irrigation areas accelerates when the value of α is higher, which implies a heavier influence of drought on irrigation infrastructure operation. Considering that large reservoirs have multi-year storage capacity, it was assumed that LSI is more resilient to drought than SSI, which is presumably under certain surface-based irrigation technology. A smaller baseline value of α was assigned to LSI in the assessment: 0.5 for LSI versus 1.0 for SSI.

The estimated actual area under irrigation in a certain year was translated into the beneficiary population in poverty using the vulnerability profiling for 2030 and by assuming 0.5 hectares (ha) of irrigated land supports one household comprising five people and according to the vulnerability shares calculated in the ARV analysis.

$$LSI \text{ suitability score } (0\sim100) = [S_1 + S_2 + S_3]/3$$

$$SSI \text{ suitability score } (0\sim100) = [S_1 + \max(S_2, S_3) + S_4 + S_5]/4$$

In what follows, starting from current cropping patterns provided in the SPAM (Spatial Production Allocation Model) database, the expansion pathways of LSI and SSI are simulated. The key assumptions made in the simulation include these:

- The occurrence of LSI precedes SSI. That is, the extent of area with LSI investment potential was first determined and was excluded in the simulation for SSI.
- Within a country, the currently existing rainfed cultivation area will be first converted to irrigated area before new area is brought into cultivation/irrigation. Furthermore, the rainfed-irrigation area conversion and the irrigation-induced farming area expansion occurs according to the suitability score determined in geographic information system mapping analysis from pixels with high environmental suitability scores to pixels with medium or low scores.
- A B/C analysis of irrigation investment was estimated in the simulation. The uptake of irrigation will not occur if the production of irrigated crops is not profitable. The calculated B/C ratio will also influence the mix of irrigated crops. Farmers tend to use irrigation to cultivate high-value crops. The area of an irrigated crop was assumed be proportional to its economic profitability.
- Irrigation expansion is also constrained by availability of water resources and demand for irrigated crop products. The irrigation expansion stops when (a) water resources available for irrigation are exhausted or (b) production of irrigated crops reaches the projected national demand by 2030. The water balance accounting was based on hydrologic and crop simulation results using IFPRI's SWAT model for Sub-Saharan Africa, and food demand in 2030 was estimated using IMPACT.

Cost Estimates

Evaluating the Costs of Resilience Interventions

Drylands development policies must take into account not only the extent to which interventions can reduce vulnerability and increase resilience, but also the cost of implementing those interventions. Since interventions considered in this book are already available "on the shelf" and are ready for implementation, research and development costs are sunk costs and can safely be ignored. Additional costs that need to be considered include these:

- Private costs associated with technology adoption (for example, the costs incurred by herders and farmers when purchasing inputs and/or hiring additional labor)
- Public costs associated with technology transfer (for example, the costs of extension campaigns and farmer training)
- Miscellaneous overhead costs.

Because technology transfer costs vary considerably depending on the accuracy of targeting, costs were estimated for three scenarios:

- *Zero targeting:* All technologies are promoted in all polygons having nonzero cropping area.
- *Intermediate targeting:* All technologies are promoted only in polygons having non-zero cropping area and in which farm-level benefits exceed technology transfer costs.
- *Perfect targeting:* Among the technologies having positive farm-level benefits, the only technology promoted is the one with the greatest impact on resilience, that is, the one producing the largest reduction in the number of drought-affected people.

Depending on the accuracy of targeting, the average annual cost across the entire sample of dryland countries ranges from US$140 million to US$1.31 billion (table 2.13). Costs on this order of magnitude compare favorably with current levels of development assistance provided in dryland countries.

Livestock Interventions Costs

Cost estimates for the analysis of livestock systems are based on cost projections from five recently launched internationally funded projects dealing with pastoral areas.[11] These data were complemented with data obtained through a review of the literature. Table 2.14 provides a summary of the cost per pastoralist/agro-pastoralist associated with these projects.

Table 2.13 Estimated Annual Costs of Resilience Interventions (US$ Billions/year)

Cost item	Zero targeting	Intermediate targeting	Perfect targeting	Other
Private—livestock and crops	1.09	0.36	0.12	
Private—irrigation				2.18
Public	0.21	0.06	0.02	
Total	1.30	0.42	0.14	

Source: World Bank calculations.
Note: Irrigation costs are reported separately because the targeting of irrigation investments is "built-in" to the analysis, which assumes that irrigation development occurs only in locations where the investment is expected to generate an internal rate of return of 12 percent or more.

Table 2.14 Average Cost/Person/Year of the Main Interventions in Five Dryland Livestock Development Projects (US$)

	Average cost/person/year (US$)	Number of projects	Range (US$)
Health improvement	3.95	3	3.37–20.12
Market improvement (early offtake of bulls)	6	3	3.67–8.33
Early warning systems	3.72	2	1.79–2.09
Social services, etc.	5.3	2	2.39–5.82

Source: De Haan 2016.

Table 2.15 Assumptions about the Allocation of Adoption- and Non-Adoption-Related Costs

Item	Allocation
Animal health non-adoption-related	Of total health improvement budget, 20% in investments and 25% in recurrent costs
Animal health adoption-related	Of total health improvement budget, 25% in investment and 30% in recurrent costs
Animal health improvement adoption-related by livestock system	10% higher/person (higher delivery costs) in pastoral systems
Early offtake (market integration)	Of total budget, 70% in investment and 30% in recurrent costs (high capital investment needed in infrastructure such as transport, processing facilities)
Early offtake non-adoption-related costs	Nil, because of its currently nascent character
Adoption rate	70% for pastoral and 80% for agro-pastoral households for health improvement and 60% and 70%, respectively, for early offtake
Public and private sector contribution	Public sector: 80% for cross-cutting costs, 60% for adoption costs in animal health improvement, and 20% for early offtake; the remainder belongs in the private sector

Source: De Haan 2016.

The range of values is significant, particularly for health improvement. However, the average is in line with the estimates of the World Organisation for Animal Health (OIE)-sponsored study for Uganda (CIVIC Consulting 2009). For development decision making, it is important to know the distribution between technology adoption-related and non-adoption-related costs, as well as between investment and recurrent costs. The assumptions used were based on the projects analyzed and the authors' experience (table 2.15).

In aggregate, these figures seem high, at a total of about US$10 billion over the 20-year period or about US$500 million/year (about US$200 million/year for the public sector). They look more reasonable when calculated per person (number of people made resilient), as shown in table 2.16.

Figure 2.10 shows that with the exception of Niger, the costs per person made resilient are significantly below the US$100 per person per year normally calculated for food aid. As expected, the annual cost per person made resilient is higher in pastoral areas. In general, the costs in East Africa seem to be lower than in the Sahel. At an average cost of US$27 per person per year, they are less than half the US$65 per person per year for Ethiopia and Kenya as estimated by Venton et al. (2012).

Table 2.16 Summary of Costs (Average 2011–14 Prices, US$ Billions) of Health and Early Offtake Interventions and Their Distribution between the Public and Private Sectors (2011–30)

	Cross-cutting costs	Adoption costs animal health	Early offtake costs	Total
Public sector	1.14	1.69	1.18	4.01
Private sector	0.29	1.13	4.71	6.12
Total	1.43	2.82	5.89	10.13

Source: De Haan 2016.

Figure 2.10 Estimated Unit Cost (US$/Person Made Resilient/Year, Expressed on a Log Scale) under Baseline Climate and Health and Early Offtake Scenarios

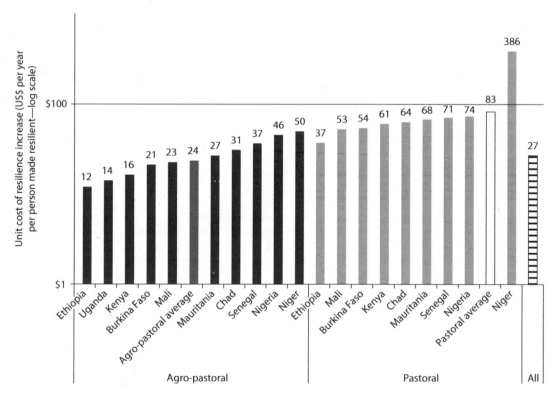

Source: De Haan 2016.

Rainfed Crop Intervention Costs

The cost of adopting the rainfed cropping technologies includes public costs borne by the public sector during an initial period when a technology is first introduced (for example, costs associated with extension campaigns, demonstrations, and free samples; see table 2.16) as well as private costs borne by the adopting farmers themselves (such as the cost of purchasing seed or fertilizer or the cost of performing additional operations such as planting fertilizer trees or building water-harvesting structures). Private costs were included in the analysis by adjusting downward the yield gain associated with adoption of the technology by a discount factor estimated to represent the cost of adopting the technology.

To reflect the fact that farm households will use part of their income to purchase the inputs required for adopting the technology (labor, seed, fertilizer), costs were expressed in terms of the crop equivalent of purchasing the required inputs, with production valued at country- and crop-specific farm gate prices, calculated as averages of the corresponding FAOSTAT values over the period 2000–12. The cost (estimated on the basis of the literature and expert judgment) varied by technology (tables 2.17 and 2.18). In some cases it was modest (such as for adoption of drought-tolerant and heat-tolerant varieties, adoption of FMNR), whereas in other cases it was more substantial (additional fertilizer, water harvesting).

In recognition that technology adoption costs may be borne by the farmer or by the state (in the form of subsidies), sensitivity analysis was carried out to explore the impacts on adoption incentives of differing levels of private costs. To reflect the fact that the best-bet crop-farming technologies will not all be profitable in every location, a switch was built into the model to determine which technology is adopted in any given polygon. The switch follows the logic of the intermediate targeting of interventions, meaning that if adoption of a given technology has the effect of reducing the number of drought-affected people in a specific administrative unit or a country, that technology is deemed effective

Table 2.17 Public Costs of Technology Transfer (US$/hectare)

Description	Millet	Sorghum	Maize
1. Drought tolerance	1.25	1.35	1.5
2. Heat tolerance	1.25	1.35	1.5
3. More fertilizer	10	10	10
4. (a) Agroforestry, 5 trees/ha	45	45	45
4. (b) Agroforestry, 10 trees/ha	45	45	45
5. Water harvesting	20	20	20

Source: IFPRI 2015.
Note: ha = hectare.

Table 2.18 Private Costs of Technology Adoption (US$/hectare)

Description	Millet	Sorghum	Maize
1. Drought tolerance	3	3	15
2. Heat tolerance	3	3	15
3. More fertilizer	30	30	30
4. (a) Agroforestry, 5 trees/ha	7	7	7
4. (b) Agroforestry, 10 trees/ha	9	9	9
5. Water harvesting	45	45	45

Source: Based on data from IFPRI 2015.
Note: ha = hectare.

and retained for that area. If, on the contrary, adoption of that technology is not able to oppose the natural increase in the number of drought-affected people in an area, the technology is deemed ineffective and is discarded. Since synergies resulting from the simultaneous adoption of multiple technologies are not captured well by the DSSAT model, the analysis used the simplifying assumption that only the technology that proves to be most effective in terms of reduction in the number of people affected is adopted in a given location (perfect targeting). Because simultaneous adoption of multiple technologies would certainly result in additional benefits (in terms of yield increases and income gains), the resilience-enhancing impacts of adoption of improved rainfed cropping technologies should be considered conservative.

Irrigation Costs

Given the considerable uncertainty and the wide range of irrigation technology and expansion costs, three sets of cost assumptions were considered in the analysis of irrigation development, ranging from US$8,000–US$30,000 per ha for LSI, and from US$3,000–US$6,000 per ha for SSI (table 2.19). The medium-cost assumptions were used for the baseline scenario.

Consolidating the Results of the Resilience Analysis

Estimated reductions in the numbers of drought-affected people likely to result from interventions in livestock systems and rainfed cropping systems, as well as from investments in irrigation, are consolidated and compared in the next section. Key elements of the consolidation process included the following:

- The livestock model was used to generate estimates of the number of vulnerable people (with and without the interventions) in hyper-arid and arid zones only (aridity classes 1–3; table 2.20), using the model's parameters for pastoral livelihoods.
- Results expressed in terms of number of households were converted into numbers of people by assuming an average household size of six people.
- The number of drought-affected people in the livestock model was estimated by applying country-specific drought incidence factors (average number of

Table 2.19 Irrigation Development Unit Cost Assumptions (US$/hectare)

	Low		Medium		High	
Cost scenario	Capital	Operation and maintenance	Capital	Operation and maintenance	Capital	Operation and maintenance
Large-scale irrigation	8,000	800	12,000	1,200	30,000	3,000
Small-scale irrigation	3,000	100	4,500	125	6,000	150

Source: Xie et al. 2015.

Table 2.20 Coverage of Livelihood Modeling by Aridity Zone

Aridity zone	Vulnerability model	Affected people
1	Livestock/Pastoral	Livestock model applying coefficients from the crop model
2	Livestock/Pastoral	Livestock model applying coefficients from the crop model
3	Livestock/Pastoral	Livestock model applying coefficients from the crop model
4	Crop	Crop
5	Crop	Crop
6	Crop	Crop

drought-affected people as percentage of vulnerable people) obtained from the crop model. This is justified given the likely significant correlation between drought impacts on the staple crops modeled (maize, millet, sorghum) and impacts on the grasses found in rangelands.

- The number of drought-affected people engaged in crop farming in aridity classes 4–6 (including both the original crop farmers and the people who are likely to transition from pastoralism to farming) was estimated using the ARV software's underlying model, adapted to use yield data as an input instead of the drought index. The approach used in this book did not consider the significant scope for implementing livestock-related interventions in agro-pastoral systems found in semi-arid and dry sub-humid zones. For this reason, while the modeling results indicate the order of magnitude of the likely resilience benefits of the different interventions, they represent conservative lower-bound estimates of the full potential.

Notes

1. SHIP (Survey-Based Harmonized Indicators Program).
2. Note that this approximation is crude: π_1 and p_m are derived from IFPRI's classification of all agricultural livelihoods across 16 livestock systems, which are aggregated into either "pastoralism" or "mixed farming," but ignores crop farmers without livestock. These ratios are then directly applied to the share of the agricultural population that was not in crop farming. It was also assumed that the ratios of the population in crop farming, in pastoralism, and in mixed farming remain constant over time.
3. http://iresearch.worldbank.org/PovcalNet/povOnDemand.aspx.
4. http://www.africanriskcapacity.org/2016/10/31/africa-riskview-methodology/.
5. A tropical livestock unit (TLU) is used to aggregate different species and age classes of livestock based on feed requirements. For this study, the following conversion factors were used: camel = 0.7 TLU; cattle = 0.6 TLU; and sheep and goats = 0.1 TLU.
6. The ranking of distributions to determine the best fit is given by the Akaike information criterion, http://en.wikipedia.org/wiki/Akaike_information_criterion.
7. http://www.africanriskcapacity.org/2016/10/31/africa-riskview-methodology/.
8. Decision Support System for Agrotechnology Transfer.

9. Myriam Adam, personal communication, August 25, 2011.

10. The adjusted soil profiles are available at https://gist.github.com/jawoo/00411e5e5d7 ced0dc6e0.

11. The Ethiopia-Drought Resilience & Sustainable Livelihood Program in the Horn of Africa (PHASE I), funded by the African Development Bank (US$48.5 million, 2012); the International Fund for Agricultural Development (IFAD)- and World Bank-funded Regional Pastoral Livelihoods Resilience Project for Kenya and Uganda (US$132 million, 2014); the World Bank-funded Regional Sahel Pastoralism Support Project (US$250 million, under preparation); the World Bank/IFAD funded Ethiopia Pastoral Community Development Project–Phase II (US$133 million, 2013); and the IFAD-funded Sudan Livestock Marketing and Resilience Program (US$119 million, under preparation).

References

CIVIC Consulting. 2009. "Systems for Animal Diseases and Zoonoses in Developing and Transition Countries." Study sponsored by OIE, World Bank, and European Union. http://www.oie.int/doc/document.php?numrec=3835503.

Conforti, P. 2011. "Looking Ahead in World Food and Agriculture: Perspectives to 2050." Food and Agriculture Organization, Rome.

D'Errico, M., and A. Zezza. 2015. "Livelihoods, Vulnerability, and Resilience in Africa's Drylands: A Profile Based on the Living Standards Measurement Study-Integrated Surveys on Agriculture." Unpublished report, World Bank, Washington, DC.

De Haan, C., ed. 2016. *Prospects for Livestock-Based Livelihoods in Africa's Drylands.* World Bank Studies. Washington, DC: World Bank.

FAO (Food and Agriculture Organization). 2013. "Ecocrop Database." FAO. http://eco-crop.fao.org.

Fox, L., C. Haines, J. Huerta Muñoz, and A. Tho. 2013. "Africa's Got Work to Do: Employment Prospects in the New Century." Working Paper 13/201, International Monetary Fund, Washington, DC.

Gerber, P. J., H. Steinfeld, B. Henderson, A. Mottet, C. Opio, J. Dijkman, A. Falcucci, and G. Tempio. 2013. *Tackling Climate Change through Livestock—A Global Assessment of Emissions and Mitigation Opportunities.* Rome: Food and Agriculture Organization.

Ham, F., and E. Filliol. 2012. "Pastoral Surveillance System and Feed Inventory in the Sahel." In *Conducting National Feed Assessments,* edited by M. B. Coughenour and H. P. S. Makkar. Rome: Food and Agriculture Organization.

HarvestChoice. 2015. "Dryland Regions of West and East Africa." International Food Policy Research Institute, Washington, DC., and University of Minnesota, St. Paul.

Hoogenboom, G., J. W. Jones, P. W. Wilkens, C. H. Porter K. J. Boote, L. A. Hunt, U. Singh, J. L. Lizaso, J. W. White, O. Uryasev, F. S. Royce, R. Ogoshi, A. J. Gijsman, G. Y. Tsuji, and J. Koo 2010. "Decision Support System for Agrotechnology Transfer (DSSAT) Version 4.5 [CD-ROM]." University of Hawaii, Honolulu.

IIASA/FAO. 2012. Global Agro-ecological Zones (GAEZ v3.0). IIASA, Laxenburg, Austria, and FAO.

Jones, J. W., G. Hoogenboom, C. H. Porter, K. J. Boote, W. D. Batchelor, L. A. Hunt, P. W. Wilkens, U. Singh, A. J. Gijsman, and J. T. Ritchie. 2003. "DSSAT Cropping System Model." *European Journal of Agronomy* 18 (3–4): 235–65.

Koo, J., and J. Dimes. 2013. "HC27 Generic Soil Profile Database." http://hdl.handle .net/1902.1/20299. International Food Policy Research Institute [Distributor] V2 [Version].

Lizaso, Jon I., Kenneth J. Boote, James W. Jones, Kassahum Tesfaye, Javier Di Matteo, Jawoo Koo, Nicola Cenacchi, and Fernando Andrade. 2015. "Improving Crop Simulation Models to Cope with Crop Responses to Drought." *Geophysical Research Abstracts* 17, EGU2015-15283, 2015. http://adsabs.harvard.edu/abs/2015 EGUGA..1715283L.

MacDonald, A. M., L. Maurice, M. R. Dobbs, H. J. Reeves, and C. A. Auton. 2012. "Relating In Situ Hydraulic Conductivity, Particle Size and Relative Density of Superficial Deposits in a Heterogeneous Catchment." *Journal of Hydrology* 130 (41): 434–35. doi:10.1016/j.jhydrol.2012.01.018.

Robinson, J., and F. Pozzi. 2011. "Mapping Supply and Demand for Animal-Source Foods to 2030." Animal Production and Health Working Paper 2, Food and Agriculture Organization, Rome.

Robinson, T. P., G. R. W. Wint, G. Conchedda, T. P. Van Boeckel, V. Ercoli, E. Palamara, G. Cinardi, L. D'Aietti, S. I. Hay, and M. Gilbert. 2014. "Mapping the Global Distribution of Livestock." *PLoS One* 9 (5): e96084. doi:10.1371/journal.pone.0096084.

Rosegrant, M. W., J. Koo, N. Cenacchi, C. Ringler, R. D. Robertson, M. Fisher, C. M. Cox, K. Garrett, N. D. Perez, and P. Sabbagh. 2014. *Food Security in a World of Natural Resource Scarcity: The Role of Agricultural Technologies.* Washington, DC: International Food Policy Research Institute (IFPRI). http://ebrary.ifpri.org/cdm/ref/collection /p15738coll2/id/128022.

Ruane, A. C., R. Goldberg, and J. Chryssanthacopoulos. 2015. "AgMIP Climate Forcing Datasets for Agricultural Modeling: Merged Products for Gap-Filling and Historical Climate Series Estimation." *Agricultural and Forest Meteorology* 200: 233–48. doi:10.1016/j.agrformet.2014.09.016.

SEDAC. 2015. "The Global Urban-Rural Mapping Project (GRUMP)." Socioeconomics and Data Applications Center (SEDAC), National Aeronautics and Space Administration (NASA), Washington, DC. http://sedac.ciesin.columbia.edu/data /collection/grump-v1.

Spevacek, A. M. 2011. "Acacia (Faidherbia) albida." USAID Knowledge Services Center (KSC), Washington, DC. http://pdf.usaid.gov/pdf_docs/pnadm071.pdf.

Venton, C. C., C. Fitzgibbon, T. Shitarek, L. Coulter, and O. Dooley. 2012. *The Economics of Early Response and Disaster Resilience: Lessons from Kenya and Ethiopia.* Economics of Resilience Final Report. https://www.gov.uk/government/uploads/system/uploads /attachment_data/file/67330/Econ-Ear-Rec-Res-Full-Report_20.pdf.

World Bank. 2007. *World Development Report 2008: Agriculture for Development.* Washington, DC: World Bank.

Xie, H., W. Anderson, N. Perez, C. Ringler, L. You, and N. Cenacchi. 2015. *Agricultural Water Management for the African Drylands South of the Sahara.* World Bank Studies. Washington, DC: World Bank.

You, L., U. Wood-Sichra, S. Fritz, Z. Guo, L. See, and J. Koo. 2014. "Spatial Production Allocation Model (SPAM) 2005 Beta Version." http://mapspam.info.

CHAPTER 3

Results

Baseline Vulnerability, 2010

The business as usual (BAU) scenario projections of the numbers of vulnerable people likely to be living in the drylands of East and West Africa in 2030 provide a convenient yardstick by which to assess the attractiveness of the various interventions that are designed to improve resilience.

Exposure

The number of people currently living in drylands is over 247 million people (table 2.4), which is 48 percent of the total population living in these regions, and 46 percent and 49 percent of the total population in East and West Africa, respectively (figure 3.1).

Sensitivity

Among the 247 million people exposed to drought by living in drylands, over 171 million are dependent on agriculture and are therefore sensitive to drought. The sensitive population accounts for over 69 percent of the total population living in the drylands and includes households generating their income from crop farming (16 percent), mixed farming (43 percent), and pastoralism (10 percent).

The pastoralist population ranges from 9 percent of the total drylands population in West Africa to 14 percent in East Africa, while crop-farming activities are the main source of income for 14 percent of drylands households in West Africa and 19 percent in East Africa. Figure 3.2 shows the differences in the composition of people sensitive to drought in East and West African countries. Tanzania, in East Africa, has the highest number of farmers (almost 8 million), while the biggest share of pastoralists lives in Sudan. In West Africa, Chad and Nigeria have the highest numbers of pastoralists and farmers, respectively.

Vulnerable and Drought-Affected People

Across the 10 dryland countries for which sufficient data were available to allow modeling of resilience interventions, it was estimated that in 2010 about 30 percent

Figure 3.1 Exposure Level in East and West Africa, 2010

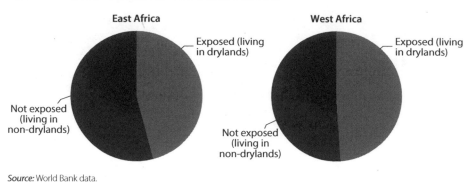

East Africa

Exposed (living in drylands)

Not exposed (living in non-drylands)

West Africa

Exposed (living in drylands)

Not exposed (living in non-drylands)

Source: World Bank data.

Figure 3.2 Estimated Drylands Population Dependent on Agriculture, by Country and Livelihood Type (Millions of People), Selected West and East African Countries, 2010

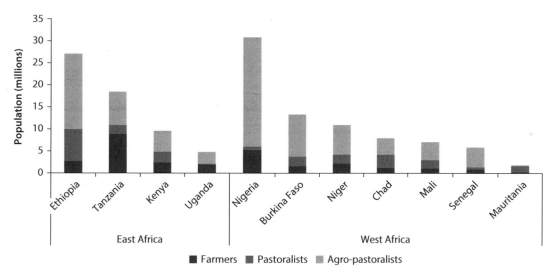

Source: World Bank calculations.

of the population living in dryland zones was vulnerable to droughts and other shocks. While this number is quite large, fortunately not all vulnerable households experience a drought every year. Assuming historical climate patterns, the modeling simulations show that in any given year, approximately 20 percent of vulnerable households are affected by drought, equivalent to about 6 percent of the total population in the 10 countries. Depending on the country, the average share of people living in drylands expected to be affected by drought in any given year ranges from 7 to 20 percent, with an overall average of 14 percent (figure 3.3).

The estimated distribution of drought impacts is shown in map 3.1. These figures are of particular importance because they determine the amount of resources that will have to be committed on a long-term basis to fund social safety nets to provide support to all of the people affected by droughts.

Figure 3.3 Percentage of People Vulnerable to and Affected by Drought, Selected West and East African Countries, 2010

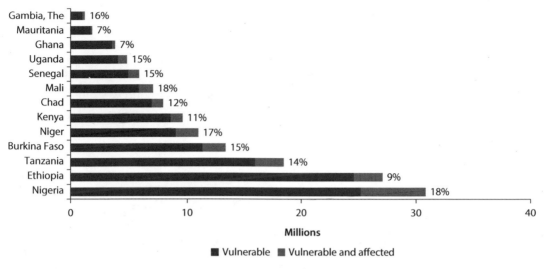

Source: World Bank calculations.
Note: The figures appearing to the right of the bars indicate the average number of drought-affected people, expressed as a percentage of the total number of vulnerable people.

Map 3.1 Projected Number of Drought-Affected People, Annual Average, Selected Countries, 2010

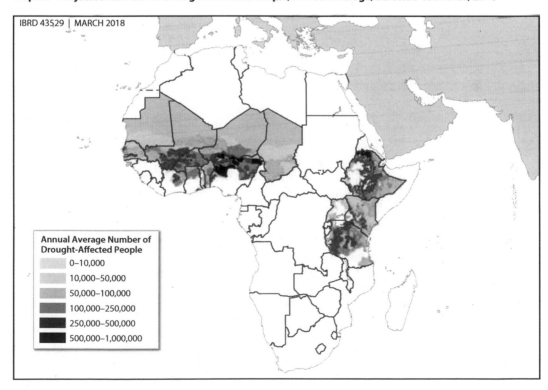

Source: Calculations based on data from African Risk Capacity Agency calculations 2015.

Mitigating Drought Impacts in Drylands • http://dx.doi.org/10.1596/978-1-4648-1226-2

Of course, these are not the same people every year, as droughts occur in different locations and strike with different intensities. The size of the drought-affected group is of interest because it determines the amount of resources to be mobilized every year—in the form of safety nets, international humanitarian assistance, or other forms of support—to assist people unable to cope with the effects of drought. The size of the drought-affected group also influences the mix of assistance that can be offered: for a given budget, the larger the group of drought-affected households, the larger the share of resources needed for short-term emergency response activities and the smaller the share of resources available to build longer-term resilience.

Baseline Vulnerability, 2030

Exposure

The number of people living in drylands who are exposed to droughts and other shocks will grow considerably by 2030. Barring an unexpected acceleration in rural-urban migration (that is, beyond the trend already built into the UN population projections), by 2030 the population living in rural areas of the dryland countries is projected to grow between 40 percent and 120 percent (figure 3.4).

Economic growth will reduce the share of people living in drylands who are sensitive to droughts and other shocks, but probably not fast enough to overcome the effects of demographic growth, as shown in the case of Niger, which has one of the highest population growth rates in the world. As gross domestic product (GDP) growth generates new employment opportunities in the manufacturing

Figure 3.4 Projected Rural Population in 2030 (2010 = 100, Medium Fertility Scenario), Selected West and East African Countries

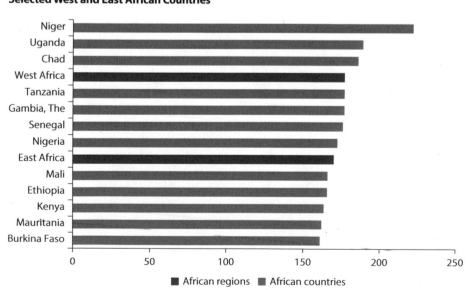

Source: United Nations World Population Prospects, 2015 Revision (http://www.un.org/en/development/desa/publications/world-population-prospects-2015-revision.html).

and services sectors, the share of the population living in drylands and dependent on livestock-keeping and crop farming is likely to decrease. Nevertheless, in the presence of rapid population growth, the absolute number of people who depend on these two predominant livelihood strategies and who are exposed and sensitive to droughts and other shocks will likely outpace the exits from agriculture. Also, even though the estimated rural-to-urban migration will decrease the number of people exposed to shocks in 2030, in many countries this will not offset other demographic changes occurring over the next 20 years.

Sensitivity

The assumption is that GDP growth by 2030 leads to structural transformation, which causes people to exit out of agriculture (World Bank 2007). Agricultural employment is then a function of economic growth, estimated for the next 20 years using data on GDP growth during the period 1980–2010. As explained previously, to accommodate uncertainty about future GDP growth, three scenarios were modeled (slow, medium, and fast). However, as noted earlier, such a relationship might not hold across Sub-Saharan Africa.

Figure 3.5 shows the increase in the agriculture dependent population compared to 2010 by country under two GDP growth scenarios. For many countries, the projected increase falls between 40 percent and 80 percent, but in a few countries it is much higher (100 percent or more for Chad and Niger).

Figure 3.5 Number of People in Drylands Projected to Be Dependent on Agriculture in 2030 (2010 = 100, Medium Fertility Scenario), Selected West and East African Countries

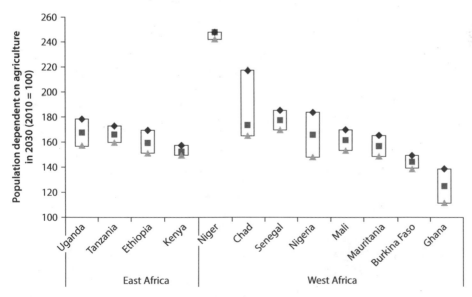

Source: World Bank calculations.

Note: The figures in the chart represent the number of dryland people projected to be dependent on agriculture in 2030 in relation to the corresponding figure in 2010. For example, a figure of 140 indicates a 40 percent increase over the 2010 level of agricultural employment. For each country, the range is defined by different scenarios of per capita GDP growth, which is expected to generate some exit of employment out of agriculture as a result of structural transformation.

With a few exceptions (Chad and Nigeria), the results are not very sensitive to the assumptions made about future GDP growth. The average increase in East Africa is around 60 percent, while in West Africa the growth rates of the agriculture-dependent population vary from 40 percent in Burkina Faso to over 140 percent in Niger.

In the best possible scenario for the drylands (low fertility rates and high GDP growth), the number of people involved in agriculture-related activities (and therefore sensitive to drought) in 2030 will increase almost 50 percent. In the worst-case scenario (low GDP growth and high fertility rates), the increase might reach 90 percent. Taking into consideration the median of all possible scenarios, the likely increase in the number of people generating income from drought-sensitive activities like agriculture and livestock is about 67 percent compared to the number of people dependent in 2010.

Looking at the composition of the people dependent on drought-sensitive activities for their livelihoods in all three GDP growth possibilities, the increase seems uniform across all livelihood activities, namely crop farming, pastoralism, and mixed farming, and to some degree proportional to the GDP growth scenario (figure 3.6). A slightly lower increase is recorded in crop-farming activities, possibly because of the reduction of arable lands in the drylands by 2030, while pastoralism and mixed farming have slightly more substantial increases across all scenarios of GDP growth.

A second group of significance for the analysis consists of pastoralist households living in arid zones who own herds smaller than the minimum size needed to provide enough income to meet household consumption needs (estimated to be 5 total livestock unit [TLU]/household). For these households, day-to-day survival appears extremely precarious, even in the absence of

Figure 3.6 Percentage Change in Agriculture-Dependent Population, 2010–30, by Livelihood System under Medium-Fertility Scenario

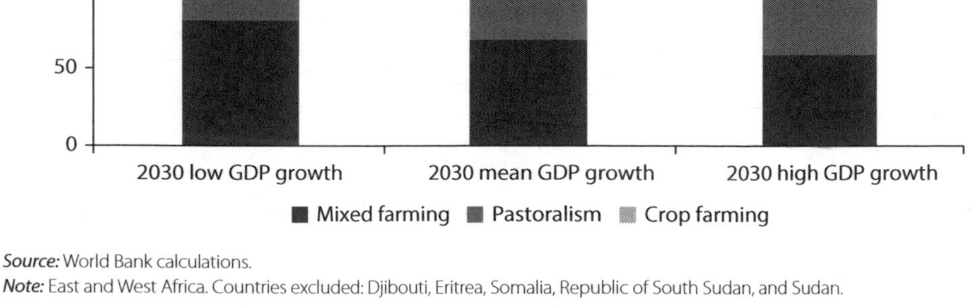

Source: World Bank calculations.
Note: East and West Africa. Countries excluded: Djibouti, Eritrea, Somalia, Republic of South Sudan, and Sudan.

Figure 3.7 Share of 2010 Population Likely to Drop Out of Pastoralism by 2030, Selected West and East African Countries

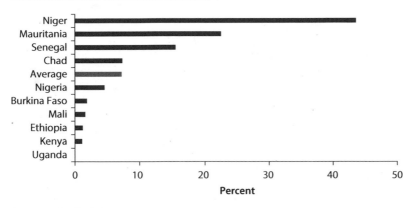

Source: World Bank calculations.

droughts and other shocks. This group—which accounts for 7 percent of the population across the entire sample of 10 countries but makes up a much larger share of the population in some countries, for example, Niger (figure 3.7), will come under increasing pressure to abandon pastoralism as its primary livelihood strategy and turn to other activities. In the model's 2030 projections, it was assumed that pastoralist households owning fewer animals than the critical minimum level of 5 TLU will transition from pastoralism to crop farming.

Vulnerable and Drought-Affected People

With the exception of Burkina Faso, by 2030 all of the countries in the sample are projected to experience increases in the number of vulnerable people. The increase is especially high in Niger, where the number of vulnerable people is expected to triple (figure 3.8). These projected increases reflect the combined effects of several key drivers, including rapid population growth, relatively slow and inequitable economic growth, and binding bioclimatic and social constraints that limit the natural resource base's ability to support greater numbers of animals. Most important, in pastoral areas, prospects for expanding herd sizes at rates fast enough to keep pace with population growth are limited by the size of accessible grazing area.

To put the magnitude of the resulting challenge in perspective, the annual cost of bringing all drought-affected people up to the poverty line by providing support through social safety nets would range from 0.3 percent to almost 5 percent of GDP (figure 3.9).

In interpreting these results it is important to keep in mind two points. First, these cost estimates are annual averages; in reality, financing needs will fluctuate dramatically and unpredictably, falling in years of normal rainfall and rising in drought years when the number of drought-affected people surges. Second, the cost estimates implicitly assume that social safety net support can be perfectly targeted to drought-affected households; in practice it is very difficult to ensure

Mitigating Drought Impacts in Drylands · http://dx.doi.org/10.1596/978-1-4648-1226-2

Figure 3.8 People Vulnerable to/Affected by Drought in 2030 (2010 = 100%)

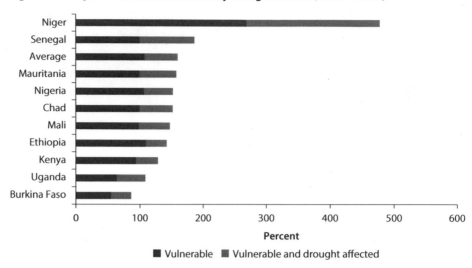

Source: World Bank calculations.

Figure 3.9 Share of 2030 GDP Required to Protect Drought-Affected Population through Social Safety Net Interventions, Selected West and East African Countries

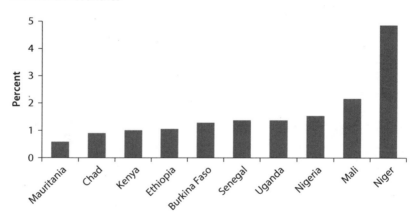

Source: World Bank calculations.

that safety net support reaches *all* drought-affected households and *only* those households. In the presence of leaks, overall financing needs would be considerably higher.

In conclusion, it is safe to assume that for most dryland countries, relying on social protection instruments to protect vulnerable populations against the effects of drought shocks will likely be beyond their fiscal means and institutional capacity.

Three scenarios were considered to explore the likely impacts of different rates of growth and different poverty-reducing effects of growth (figure 3.10). A pessimistic, low-end scenario assumes that growth will be

Figure 3.10 Vulnerable People in Drylands in 2030 (2010 = 100, Medium-Fertility Scenario), Selected West and East African Countries

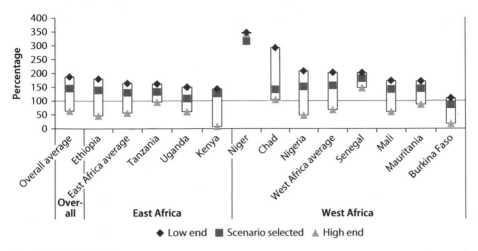

Source: World Bank calculations.

slow and non pro-poor. An optimistic, high-end scenario assumes that growth will be rapid and pro-poor. An intermediate scenario (used for the rest of the analysis) assumes that growth will be moderate and that the poverty-reducing effect will be modest (growth elasticity of poverty reduction [GEPR] = 0.75). In most countries in East and West Africa, only under the high-end scenario does the number of poor people decrease (signifying an increase in the ability to cope with the effects of drought and other shocks). This result is not universal, however; Niger and Chad are notable exceptions. Under the intermediate scenario, the number of poor people increases significantly (signifying a decrease in the ability to cope with the effects of drought and other shocks). Across the entire set of countries, the number of vulnerable people increases by 45 percent. The increase is smaller in East Africa (40 percent) compared to West Africa (55 percent). The increase is particularly high in Senegal (80 percent) and Niger (200 percent).

Investment in girls' education can mitigate but not fully address the vulnerability challenge. Investment in the education of girls has been shown to lower fertility rates over the medium to long term (Summers 1992; UNESCO 2011). However, as fertility rates fall, so does the number of people who are likely to need access to safety nets. The impact of reducing fertility rates, while non-negligible, is likely to be limited.

Using the United Nations (UN) low fertility population projections as a first-order approximation of the effects of fertility reduction policies, the increase by 2030 in the number of people vulnerable to shocks could be reduced by 30–45 percent (figure 3.11).

These sobering results underline the enormity of the challenge facing African governments and the development community more widely. They point to the importance of assessing the ability of different types of interventions to increase the resilience of the poorer segments of the dryland population.

Figure 3.11 Vulnerable People in Drylands in 2030 (2010 = 100, Different Fertility Scenarios), Selected West and East African Countries

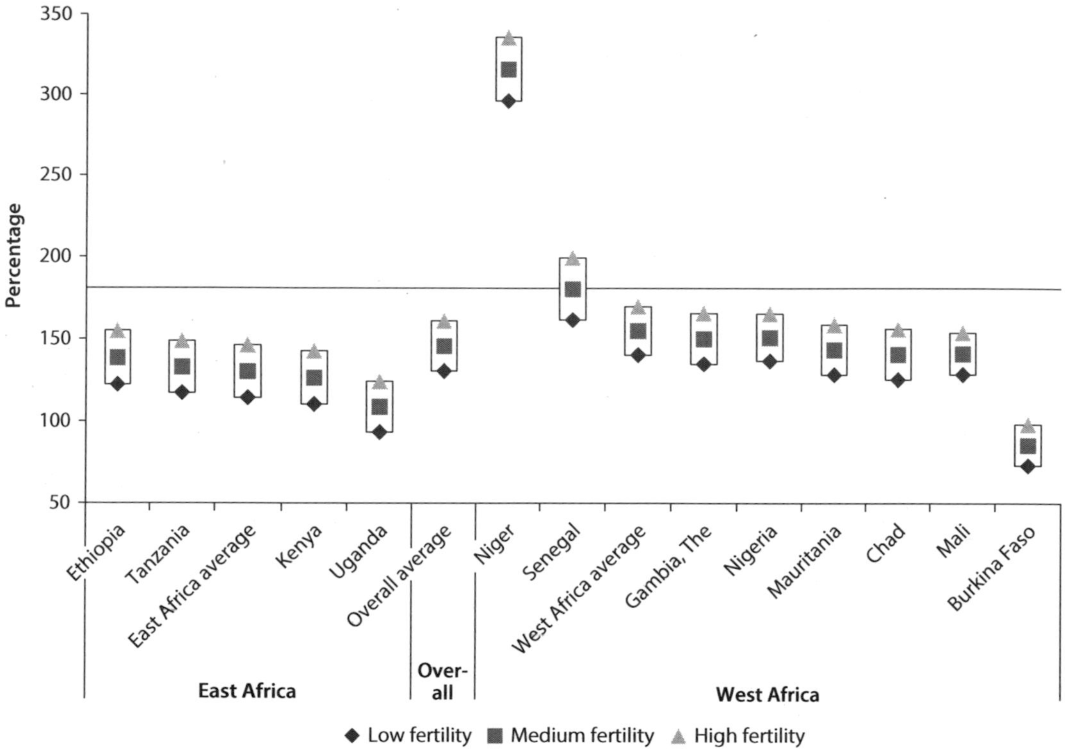

◆ Low fertility ■ Medium fertility ▲ High fertility

Source: World Bank calculations.

Intervention Results

Livestock

The modeling covers for the pastoral sector the improvement of animal health services, and integration of dryland areas with the higher rainfall areas of the sub-humid and humid tropics. The main effect of the animal health measures is a reduction in mortality, particularly of young stock. Over a 20-year period, the mortality of young stock is estimated to decrease by 35 percent in cattle and 40 percent in sheep and goats. The main impact of early offtake of young adult livestock is an increase in price (estimated at 10 percent per kilogram live-weight) and a decrease in the risk for the herder of financial losses from low sale prices. The effect of these livestock interventions on the number of TLU needed to attain the income to become resilient is estimated through ECO-RUM herd modelling, based on the information sources described in figure 2.3 and shown in table 3.1.

For the baseline weather and the severe drought scenarios, through the ECO-RUM model, the number of TLU corresponding to the different vulnerability categories was then estimated for pastoral households.

Mitigating Drought Impacts in Drylands • http://dx.doi.org/10.1596/978-1-4648-1226-2

Table 3.1 Minimum Number of TLU per Household Required to Attain Resilience

| Intervention | Baseline weather | | | Mild drought | | | Severe drought | | |
	None	Health improvement	Health improvement + early offtake of male cattle	None	Health improvement	Health improvement + early offtake of male cattle	None	Health improvement	Health improvement + early offtake of male cattle
Pastoralists	14.8	11.5	11.0	16.3	12.7	12.0	17.4	13.6	13.1
Agro-pastoralists	9.0	5.3	5.2	9.9	5.9	5.8	13.1	6.0	5.9

Source: De Haan 2016.
Note: TLU = total livestock unit.

Figure 3.12 The Effect of the Two Key Interventions on Vulnerability Levels under the Baseline Weather Scenario

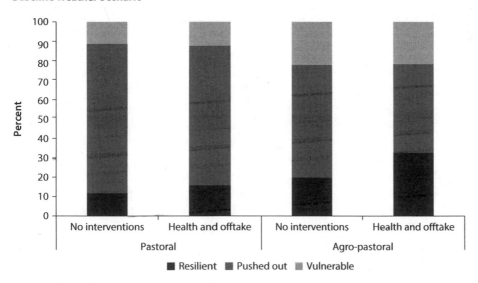

Source: De Haan 2016.

In pastoral systems, the improvements lead to only a 5 percentage point decrease in the number of households likely to exit, compensated by an increase in the share of resilient households (figure 3.12).

Beyond the model, a number of other scenarios were also tested (see De Haan 2016). Such scenarios address the major inequity found in the Survey-Based Harmonized Indicators Program (SHIP) survey data and confirmed in the literature. The proposed actions include the followiing:

- *Increasing feed production* by (a) opening up unused land, currently inaccessible because of watering constraints, and (b) safeguarding herd mobility; and
- *Consolidating pasture land* by (a) maintaining the area allocated to resilient households constant at the expected 2030 level and (b) allocating the remaining area exclusively to vulnerable households.

Policy measures to promote consolidation of pastureland include the following:

- *Stopping land grabbing* by large herd owners and enhancing mobility
- *Allocating exclusive water and grazing rights* for the wet and dry season to groups of vulnerable smallholder livestock keepers
- *Directly focusing on redressing inequities,* for example, by the following:
  - Progressive taxation of large herd owners, either through a direct tax per head or progressive grazing and watering fees
  - Differential service fees (such as for vaccination) for large herd owners
  - Introduction of, or an increase in, the export tax, as large herd owners supply more animals for export
- *Promoting small ruminants* (sheep and goats) production
- *Introducing measures* (training, microcredit, and infrastructure) to open alternative sources of income.

Simulations in De Haan 2016 suggest that only a complete package of measures would reduce vulnerability in a major fashion.

As can be seen, the technological interventions increase the share of resilient households and reduce the number of people that are likely to be forced out. But particularly for pastoral households, the impact is rather limited. For agro-pastoral households, the impact is more substantial because of higher initial disease and mortality levels. According to these simulations, however, large groups of pastoral and agro-pastoral households will still be in absolute poverty or will be pushed out.

More substantial and far-reaching technological and policy interventions were thus simulated, despite acknowledging that the policy options especially will face major implementation challenges. Their cumulative effects are provided in figure 3.13.

Figure 3.13 shows that only a concerted effort, combining technological and policy interventions, will substantially improve the outcomes of the livestock-keeping population of the drylands.

At the macro level, the effects of technological interventions on production (offtake and herd growth) are provided in table 3.2.

As can be expected, the effect on cattle is major; under severe drought, cattle experience a significant decline. Improvement in the animal health situation can reduce losses.

In pastoral areas, where only livestock-related interventions were considered (specifically, improved animal health services and early offtake of young male animals), the most important benefit is to slow the exit of the poorest herders who otherwise would be forced to abandon pastoralism and take up other livelihood activities (mainly crop farming).

By increasing livestock productivity, the livestock-related interventions reduce the minimum number of TLU needed to generate the amount of income required by livestock-dependent households to remain above the poverty line. In this way, the livestock-related best-bet interventions are

Figure 3.13 The Cumulative Effect of Key Interventions on Vulnerability Levels under the Baseline Weather Scenario

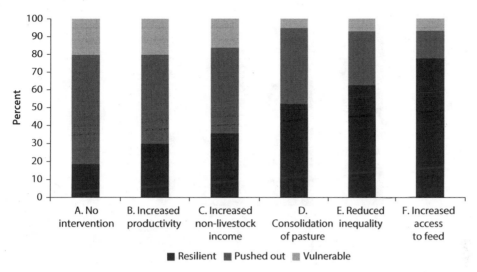

Source: De Haan 2016.

Table 3.2 Livestock Population Growth (Offtake + Population Growth), 2012–30, as Affected by Technology and Climate

	West Africa			East Africa		
	Cattle	Goats	Sheep	Cattle	Goats	Sheep
Climate scenario	(%)	(%)	(%)	(%)	(%)	(%)
Baseline	23	42	43	10	34	34
Mild drought	7	11	13	−5	10	10
Severe drought	−7	11	10	−17	9	7
Improved animal health services (mild drought)	9	36	29	10	20	12
Early offtake of young males for fattening	3	n.a.	n.a.	6	n.a.	n.a.

Source: CIRAD/MMAGE model.
Note: n.a. = not applicable.

projected to reduce the number of exits from pastoralism by 6 percent on average.

The effect is much higher in some countries, such as Kenya (13 percent fewer exits), Burkina Faso (14 percent fewer exits), Mali (16 percent fewer exits), and Ethiopia (19 percent fewer exits) (figure 3.14).

Rainfed Crops

With respect to crop-farming interventions, the biggest influence on reducing the number of drought-affected people is projected to come from improvements in soil fertility management, followed by irrigation development, adoption of drought-resistant varieties, and uptake of tree-based systems. The benefits of the

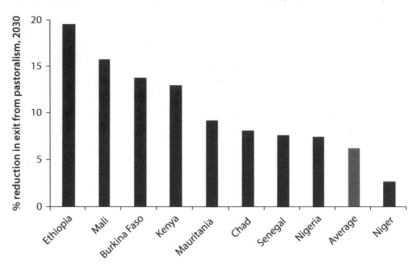

Figure 3.14 Reduction in Exits from Pastoralism because of Technological Interventions, Selected West and East African Countries, 2030

Source: World Bank calculations.

Figure 3.15 Relative Contributions of Technological Interventions in Reducing Vulnerabllity, Selected West and East African Countries, 2030

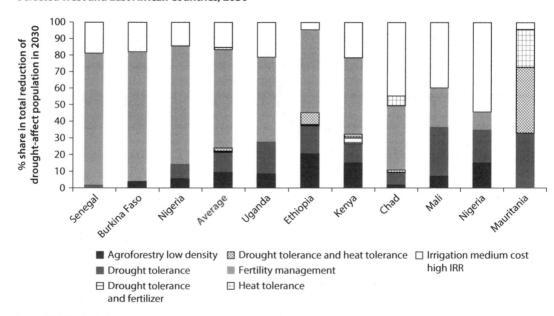

Source: World Bank calculations.
Note: IRR = internal rate of return.

Figure 3.16 Relative Contributions of Technological Interventions in Reducing Vulnerability, by Aridity Zone, 2030

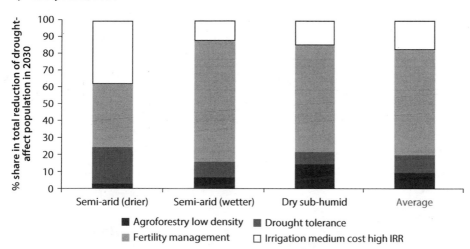

Source: World Bank calculations.
Note: IRR = internal rate of return.

different crop-farming interventions vary considerably by location, and the mix of optimal interventions is quite variable across countries. This points to the importance of carrying out location-specific assessments and tailoring interventions to meet local circumstances (figure 3.15).

Not surprisingly, the mix of optimal crop-farming interventions varies by aridity zone (figure 3.16).

In the drier parts of the semi-arid zone (Aridity Index 0.2–0.35), irrigation development is likely to have the largest impact, followed by adoption of soil fertility management practices and drought-tolerant varieties. In the wetter parts of the semi-arid zone and the dry sub-humid zone (Aridity Index 0.36–0.65), adoption of soil fertility management practices is likely to have the biggest impact by far. Adoption of tree-based systems/farmer managed natural regeneration (FMNR) is likely to have a larger impact in the dry sub-humid zone compared to more arid zones.

One important positive message emerging from the modeling work is that when accurately targeted, the best-bet crop-farming interventions have considerable potential to reduce the impacts of droughts.

As already clarified, accurate targeting was ensured in the modeling work by restricting implementation of the interventions only to locations in which their adoption was determined to be profitable (that is, simulated yield gains remain positive after yields are adjusted downward to reflect the cost of adopting the technology).

A second important message emerging—admittedly less positive—is that it is critically important to accurately target crop-farming interventions to locations where they will have maximum impact. The importance of accurate targeting

Figure 3.17 Importance of Targeting Technological Interventions

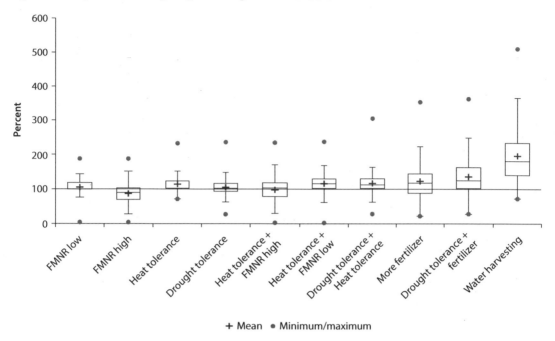

+ Mean   • Minimum/maximum

Source: World Bank calculations.
Note: FMNR = farmer managed natural regeneration.

was determined by rerunning the model and allowing the interventions to be implemented in all locations regardless of profitability. The results of this second set of simulations (summarized in figure 3.17) make clear that the cost of inaccurate targeting can be high.

The Y-axis values in figure 3.17 indicate for each technology the distribution across polygons (intersections of subnational administrative units and aridity zones) of the number of drought-affected people, expressed as a percentage of the BAU scenario. Values above 100 percent indicate poorer performance than BAU (more drought-affected people), suggesting that in the corresponding areas it is better not to adopt the technology; values below 100 percent indicate better performance than BAU (fewer drought-affected people), suggesting that in those areas it makes sense to adopt the technology. The larger the portion of the box above the 100 percent line, the larger the chance that the corresponding technology will result in an increase of the average annual number of drought-affected people.

Many of the best-bet crop-farming interventions are expected to reduce the number of drought-affected people only in selected locations. In many other locations, the cost of adopting the crop-farming intervention does not justify the expected benefits, resulting in a net loss in income and leaving adopting households more likely to be adversely affected by droughts. This means that careful assessments are needed to ensure that the best-bet crop-farming interventions are promoted only in locations in which they will actually deliver benefits by increasing resilience to drought shocks.

Tree Systems

Farmer managed natural regeneration, a set of practices farmers use to foster the growth of indigenous trees on agricultural land, has drawn substantial attention as a contributing factor to a trend of increasing vegetation greenness. FMNR is based on the regeneration of native trees and shrubs from mature root systems of previously cleared desert shrubs and trees. Regeneration techniques are used in agricultural cropland and to manage trees as part of a farm enterprise. Trees are trimmed and pruned to maximize harvests while promoting optimal growing conditions (such as access to water and sunlight).

As explained above, the area within each field that actually benefits from the decomposition of tree-contributed organic matter is determined by the canopy coverage. This was assumed to be 10 percent and 20 percent, respectively, for low-density trees (5 trees/ha) and high-density trees (10 trees/ha).

In addition to reducing sensitivity to shocks, trees can enhance the capacity of households in drylands to cope with the effects of shocks after the shocks have occurred. Trees are also assets that can be cut and sold for cash or exchanged for goods in times of need.

Crop modeling carried out for this study helped provide orders of magnitude of the benefits of FMNR in terms of reduction of drought impacts. When FMNR of native species is added to the other productivity-enhancing technologies discussed in this book, the effects are impressive. In a group of 10 countries in East and West Africa, the projected number of poor, drought-affected people living in drylands in 2030 fell—compared to the BAU scenario—by 13 percent with low-density tree systems and by more than 50 percent with high-density tree systems (figure 3.18).

Figure 3.18 Estimated Reduction in the Average Number of Drought-Affected People through Use of FMNR and Other Technologies by 2030

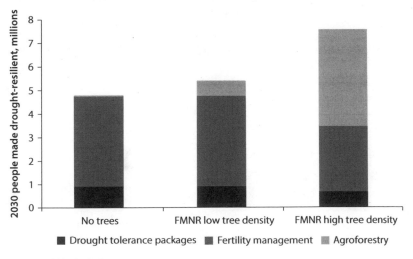

Source: World Bank calculations.
Note: FMNR = farmer managed natural regeneration.

Irrigation

Using the methodology described in the previous section, substantial potential exists to expand irrigation in Sub-Saharan African dryland area by 2030. However, the magnitude of the estimated development potential varies depending on related capital cost and assumed acceptable levels of internal rates of return (IRRs). Figure 3.19 shows how alternative capital costs of technologies and IRRs influence final irrigated area potential. The estimated land area with irrigation ranges from 0.83 million hectares to 0.39 million hectares in the three East African study countries and from 4.1 million hectares to 9.4 million hectares in the seven West African study countries. The highest irrigation potential estimate was obtained under the low-cost/5 percent IRR scenario; area declines as capital costs and IRRs increase.

Table 3.3 presents the breakdown by country and by technology under the medium-cost/5 percent IRR scenario, which is used as a baseline. The subtotal of the three East African countries is 0.7 million ha, while the subtotal of the seven West African countries amounts to 2.5 million ha. Nigeria has the largest potential for irrigation expansion: it alone holds 1.6 million ha of potentially irrigable land, almost one-half of the combined total of the 10 study countries.

As with potentially irrigable area, the size of the beneficiary population varies by irrigation costs and IRR (figure 3.20). The variation follows the same pattern: under the low-cost/5 percent IRR scenario, 0.3 million people will benefit from irrigation development and be protected against drought effects from climate variability in the three East African countries; the number is 2.3 million people in the seven West African countries. With increases in irrigation costs and acceptable level of IRR, the number of beneficiaries drops, for example to 0.1 million and 0.7 million in East and West Africa, respectively, under the high-cost/12 percent IRR scenario.

The estimated beneficiary populations under the baseline scenario fall in the middle range. Irrigation development will benefit an expected total of

Figure 3.19 Land Area with Irrigation Investment Potential under Alternative Assumptions of Irrigation Costs and IRRs

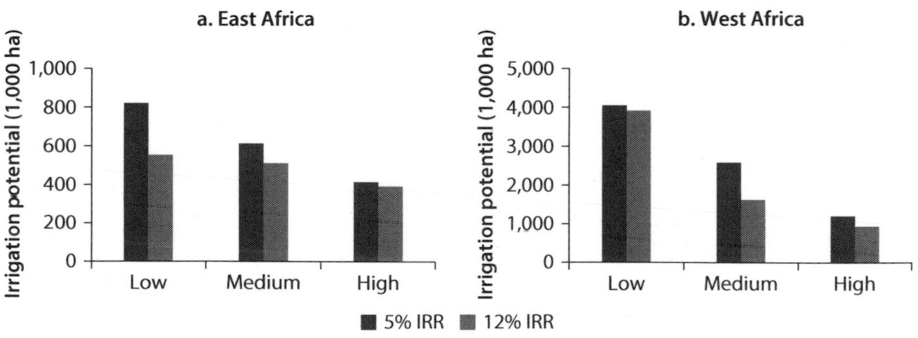

Source: Xie et al. 2015.
Note: ha = hectare; IRR = internal rate of return.

Table 3.3 Irrigation Development Potential by 2030 (1,000 ha, Medium-Cost-5% IRR), Selected West and East African Countries

Country	Large-scale irrigation potential	Small-scale irrigation potential	Total
East Africa			
Ethiopia	59	186	245
Kenya	81	255	336
Uganda	25	96	121
Subtotal	165	537	702
West Africa			
Burkina Faso	20	155	175
Chad	5	89	94
Mali	27	114	141
Mauritania	4	96	100
Niger	92	27	119
Nigeria	302	1,316	1,618
Senegal	n.a.	256	256
Subtotal	450	2,053	2,503

Source: You, Wood, and Wood-Sichra. 2009; Xie et al. 2015.
Note: n.a. = not applicable.; ha = hectare; IRR = internal rate of return.

Figure 3.20 Beneficiary Population under Alternative Assumptions of Irrigation Costs and IRRs

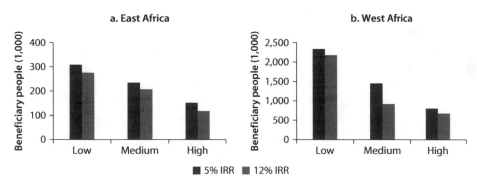

Source: Xie et al. 2015.
Note: IRR = internal rate of return.

1.7 million people: 0.2 million in East African countries and 1.5 million people in West African countries (table 3.4). Nigeria has the largest beneficiary population: 0.8 million Nigerians would benefit from irrigation.

Table 3.5 presents the size of the beneficiary population by aridity zone. Most people (1.2 million) benefiting from irrigation development live in the semi-arid (wetter) zone, which accounts for 70 percent of the total beneficiary population, followed by the dry sub-humid zone (4.5 million or 26 percent) and the semi-arid (drier) zone.

Table 3.4 Estimated Beneficiary Population (1,000 People, Medium-Cost 5% IRR), Selected West and East African Countries

East Africa		West Africa	
Country	Beneficiary population	Country	Beneficiary population
Ethiopia	86	Burkina Faso	70
Kenya	98	Chad	244
Uganda	51	Mali	120
		Mauritania	2
		Niger	120
		Nigeria	859
		Senegal	39
Subtotal	235	Subtotal	1,454

Source: Xie et al. 2015.
Note: IRR = internal rate of return.

Table 3.5 Beneficiary Population by Aridity Zone (1,000 People, Medium-Cost-5% IRR)

Aridity zone	Beneficiary population (1,000 people)
Semi-arid (drier)	74
Semi-arid (wetter)	1,170
Dry sub-humid	445
Total	1,689

Source: Xie et al. 2015.
Note: IRR = internal rate of return.

Consolidated Results

The model results show that it would be prohibitively expensive for governments in dryland countries to rely on social safety nets to protect vulnerable households from the adverse effects of droughts and other shocks. In that context, policy makers will want to know the extent to which the coming challenge can be mitigated by making current livelihood strategies more resilient. To help determine the answer, a set of interventions was selected from among the many resilience-enhancing interventions reviewed in previous sections (table 2.8). The model was used to assess the extent to which these selected interventions would reduce vulnerability among populations living in drylands by improving the productivity and sustainability of current livelihood strategies. Because of technical limitations in the model, which does not have the capacity to capture complex interactions that occur when multiple interventions are implemented simultaneously, only livestock-related interventions were considered in hyper-arid and arid zones, and only crop farming–related interventions were considered in semi-arid and dry sub-humid zones. This approach ignores the significant scope for implementing livestock-related interventions in agro-pastoral systems found in semi-arid and dry sub-humid zones. For this reason, while the modeling results indicate the order of magnitude of the likely resilience benefits of the different interventions, they represent conservative lower-bound estimates of the full potential.

Figure 3.21 Contribution of Technological Interventions to Resilience in 2030 (2010 = 100%), Selected West and East African Countries

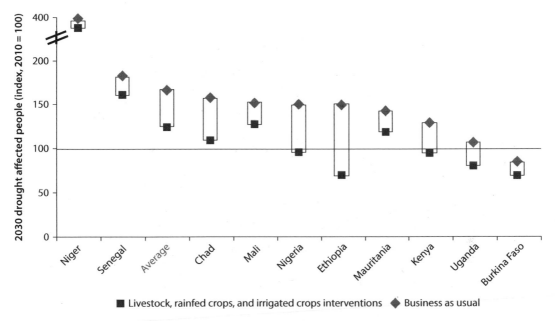

Source: World Bank calculations.

Note: The vertical axis in the figure was trimmed to accommodate Niger, an extreme outlier. The number of drought-affected farmers was estimated using the ARC (Africa Risk Capacity) model, based on the yields obtained for a set of reference staple crops (maize, sorghum, millet) grown with and without the interventions, and evaluating the number of years (over a 20-year simulated time series reflecting historical climate) in which yields fall below a certain threshold. The number of drought-affected herders was estimated based on the number of households that are able to sustain—for a given amount of biomass determined by historical climate patterns—a minimum herd size; herders lacking the minimum herd size were assumed to take up crop farming (which may or may not have made them resilient).

Results are presented for each individual intervention type and then consolidated for all aridity areas of the countries included in the analysis. The results suggest that by improving the productivity of livestock and crop-farming systems in the drylands, these interventions could considerably slow the projected increase in the number of drought-affected people (figure 3.21).

Without the interventions, by 2030 the number of drought-affected people is projected to increase by 60 percent compared to 2010. With the interventions, the number of drought-affected people is projected to increase by only 27 percent (an improvement of 43 percentage points). In some countries, notably Ethiopia and to a lesser extent Kenya and Nigeria, by 2030 adoption of improved management of livestock and crop-farming systems could reduce the absolute number of drought-affected people relative to the 2010 baseline. In other countries, particularly Niger but also Senegal and Mauritania, the best-bet interventions would have a more modest impact, and the number of drought-affected people in 2030 would still be considerably larger than in 2010.

Do Investments in Resilience Pay Off?

How cost-effective are these interventions compared to alternative strategies for reducing vulnerability and increasing resilience in the drylands? To answer this question, a simple benefit/cost (B/C) assessment was carried out. Benefits were measured in terms of reduced cash transfers needed to bring all drought-affected people up to the poverty line. The B/C analysis assumed the following:

1. In the nonintervention scenario, the number of drought-affected people would increase linearly over 15 years, as would the corresponding increase in cash transfers needed to lift them out of poverty; meanwhile, no expenditure would be made in the best-bet interventions (table 2.14).
2. In the intervention scenario, the cash transfers needed to lift drought-affected people out of poverty would increase more slowly, commensurate with the slower increase in their number. Meanwhile, public investment in the best-bet interventions would increase linearly, with the cumulative expenditure over the 15 years equaling the sum of the annual averages. The total public investment was calculated as the sum of the technology transfer cost and the overhead cost, plus 25 percent of the private cost (representing subsidies needed to encourage adoption).
3. In the intervention scenario, intermediate targeting was assumed; this implies that public agencies will be able to prescreen investments and avoid promoting technologies that are poorly suited to local agro-climatic circumstances,

Figure 3.22 Benefit/Cost Ratios of Resilience Interventions, Selected West and East African Countries

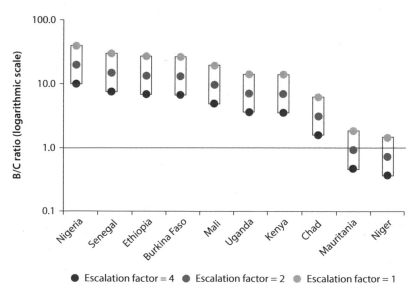

● Escalation factor = 4   ● Escalation factor = 2   ● Escalation factor = 1

Note: B/C ratios above 1 (the horizontal line on the chart) indicate that the benefits of resilience interventions exceed the costs.

but they will lack the ability to identify and exclusively promote the best-performing technology in any given location.

4. In the intervention scenario, a cost-escalation factor was used to carry out a sensitivity analysis in recognition of the fact that technology transfer costs were crude estimates and could change significantly in the future; the cost-escalation factor varies from 1 (no cost escalation) to 4 (a fourfold increase in technology transfer costs).

5. In both scenarios, the discount rate was set at 10 percent.

The results of the B/C assessment suggest that the benefits far exceed the costs of implementing the best-bet interventions (figure 3.22). In most countries (except Mauritania and Niger), the results are robust under a wide range of cost assumptions: even if costs increase fourfold, the B/C ratio remains well above 1.

References

De Haan, C., ed. 2016. *Prospects for Livestock-Based Livelihoods in Africa's Drylands.* World Bank Studies. Washington, DC: World Bank.

Summers, L. H. 1992. "Investing in All the People." Policy Research Working Paper, World Bank, Washington, DC.

UNESCO (United Nations Educational, Scientific, and Cultural Organization). 2011. *Education Counts: Towards the Millennium Development Goals.* Paris: UNESCO. http://unesdoc.unesco.org/images/0019/001902/190214e.pdf.

World Bank. 2007. *World Development Report 2008: Agriculture for Development.* Washington, DC: World Bank.

You, L., U. Wood-Sichra, S. Fritz, Z. Guo, L. See, and J. Koo. 2014. Spatial Production Allocation Model (SPAM) 2005, Beta Version. 2015. http://mapspam.info.

CHAPTER 4

Policy Implications

Enhancing the resilience of people living in the drylands will require a combination of interventions to improve current livelihoods and interventions to strengthen safety nets. An overarching recommendation emerging from the analysis reported in this book is that policy makers in dryland countries and their partners in the development community may want to look more closely at each of the two types of interventions, to assess their potential in more detail than has been possible here, taking into account local circumstances and development priorities. The Country Programming Framework prepared in the aftermath of the 2011 drought by the countries of the Horn of Africa is an important step in that direction.

Strategic plans formulated at the country level and at the regional level should be updated regularly and broadened and deepened as new knowledge becomes available, focusing especially on the medium to long term and quantifying to the extent possible the technical and financial potential of alternative interventions. With respect to the two types of interventions, this book presents detailed recommendations, which are summarized in the box 4.1 (see Cervigni and Morris [2016] for a full context). Improving current livelihood activities and strengthening social protection programs have significant potential to reduce vulnerability and enhance the resilience of populations living in drylands, but both strategies are likely to face limits. The scenario analysis carried out using the model shows that even if current livelihood strategies can be improved and social protection programs strengthened, significant numbers of households will remain vulnerable to droughts and other shocks while lacking the resources to cope effectively when a drought strikes. For these households, policy makers will need to devise strategies to facilitate the transition to alternative livelihood activities.

While the results of the modeling exercise help in defining the extent to which alternative livelihood strategies will be needed, this book does not present detailed analysis of the policy reforms and the complementary investments in human and physical capital that will be needed to help poor and vulnerable households in the drylands transition out of natural resource–based livelihoods to

Box 4.1 Selected Recommendations to Make Current Livelihoods More Resilient

(1) Livestock

- Increase production of meat, milk, and hides in drylands by developing sustainable delivery systems for animal health, promoting increased market integration, and exploiting complementarities between drylands and higher rainfall areas.
- Enhance the mobility of herds by expanding and ensuring adequate and equitable year-round access to grazing and water and by improving security in pastoral zones.
- Develop livestock early warning systems and early response systems to reduce the adverse impacts of shocks.
- Identify additional and alternative livelihood strategies, including through systems of payment for environmental services.

(2) Farming

- Accelerate the rate of varietal turnover and increase availability of hybrids.
- Improve soil fertility management.
- Improve agricultural water management.
- Promote the development of irrigation, including (a) rehabilitating existing capacity and expanding up to the viable potential (a maximum of about 10 million more hectares) and (b) focusing on small-scale systems, with good access to markets for cash crops.

(3) Natural resource management

- Promote farmer managed natural regeneration natural regeneration to establish a range of beneficial trees throughout the drylands.
- Invest in tree germplasm multiplication and promote planting of location appropriate high-value species, especially in dry sub-humid areas.
- Develop opportunities to add value to tree products produced in the drylands.

(4) Social protection

- Establish and gradually expand the coverage of national adaptive safety net programs that promote resilience of the poorest people.
- Use social protection programs to build capacity of vulnerable households to climb out of poverty, but to maintain the ability to provide humanitarian assistance in the short run.
- Respond to emergencies by scaling up existing programs, rather than relying on appeals for humanitarian assistance.
- Tailor social protection programs to address the unique circumstances of dryland populations.

productive employment in other sectors, nor does it make specific recommendations relating to these policy reforms and investments. These types of interventions fall outside the scope of the present inquiry, and further work will be needed to cover them adequately.

Reference

Cervigni, R., and M. Morris. 2016. *Confronting Drought in Africa's Drylands: Opportunities for Enhancing Resilience*. World Bank Studies. Washington, DC: World Bank.